Wind Power in Europe

Energy, Climate and the Environment Series

Series Editor: **David Elliott**, Professor of Technology, Open University, UK

Titles include:

David Elliott (*editor*)
NUCLEAR OR NOT?
Does Nuclear Power Have a Place in a Sustainable Future?

David Elliott (*editor*)
SUSTAINABLE ENERGY
Opportunities and Limitation

Joseph Szarka
WIND POWER IN EUROPE
Politics, Business and Society

Energy, Climate and the Environment
Series Standing Order ISBN 0–2300–0800–3

You can receive future titles in this series as they are published by placing a standing order. Please contact your bookseller or, in case of difficulty, write to us at the address below with your name and address, the title of the series and the ISBN quoted above.

Customer Services Department, Macmillan Distribution Ltd, Houndmills, Basingstoke, Hampshire RG21 6XS, England

Wind Power in Europe

Politics, Business and Society

Joseph Szarka
Reader in European Studies, University of Bath, UK

First published 2007 by
PALGRAVE MACMILLAN
Houndmills, Basingstoke, Hampshire RG21 6XS and
175 Fifth Avenue, New York, N.Y. 10010
Companies and representatives throughout the world

PALGRAVE MACMILLAN is the global academic imprint of the Palgrave Macmillan division of St. Martin's Press, LLC and of Palgrave Macmillan Ltd. Macmillan® is a registered trademark in the United States, United Kingdom and other countries. Palgrave is a registered trademark in the European Union and other countries.

ISBN 13: 978–1–4039–8985–7   hardback
ISBN 10: 1–4039–8985–0   hardback

This book is printed on paper suitable for recycling and made from fully managed and sustained forest sources. Logging, pulping and manufacturing processes are expected to conform to the environmental regulations of the country of origin.

A catalogue record for this book is available from the British Library.

Library of Congress Cataloging-in-Publication Data

Szarka, Joseph.
    Wind Power in Europe : politics, business and society / Joseph Szarka.
        p. cm.
    Includes bibliographical references and index.
    ISBN-1-4039–8985–0 (alk. paper)
    1. Wind Power–Europe. 2. Energy policy–Europe. 3. Renewable energy sources–Europe. I. Title.

TJ820.S98 2007
333.9'2094–dc22   2007016446

10   9   8   7   6   5   4   3   2   1
16   15   14   13   12   11   10   09   08   07

Printed and bound in Great Britain by
Antony Rowe Ltd, Chippenham and Eastbourne

Contents

List of Tables   vi

Acknowledgements   vii

Series Editor Preface   viii

List of Abbreviations   x

1   Contextualising the Wind Power Debate   1

2   Diagnosing the Wind Sector   22

3   Mobilising for Wind Power   46

4   Promoting Wind Power through National Policies   62

5   Drawing Policy Lessons from Cross-National Comparisons   88

6   Integrating Wind Power into the Electricity Supply Industry   110

7   Siting, Planning and Acceptability   138

8   Contesting Wind Power   161

9   Reviewing the Outcomes: Policy Learning and Path Choices   182

Notes   200

References   205

Index   222

List of Tables

1.1   Comparing 'hard' and 'soft' paths   4

2.1   New wind power capacity in world markets in 2005   23

2.2   Cumulative wind power capacity in the ten largest markets   23

2.3   Capacity growth of wind power in the five reference countries   24

2.4   Global market shares of wind turbine manufacturers in 2005   24

2.5   Market share of Spanish wind power developers in 2005   34

2.6   The leading operators of wind power in 2005   41

2.7   Linking scale of wind power deployment with investment and ownership models   42

4.1   RES-E targets in the European Renewables Directive   65

5.1   Extrapolation of RES-E targets contained in the European Renewables Directive   94

5.2   Prices of wind-generated electricity, 2004–5   98

6.1   Gross electricity production by fuel in percentages in 2004   111

6.2   Gross electricity production by fuel in TWh in 2004   112

9.1   Summarising the political challenges of energy sourcing   197

Acknowledgements

A very large number of people have helped the progress of this research project, some in a professional and some in a personal capacity. I am very grateful for all their help and support. In particular, I would like to thank interview respondents in Denmark, France, Germany, Spain and the UK for generously sharing their time and their knowledge. I wish to express my gratitude to the British Academy and to the Anglo-German Foundation who provided the research grants that helped make the fieldwork possible. I would like to thank the University of Bath and the Department of European Studies for giving me study leave to bring this project to fruition. I am grateful to Ingolfur Blühdorn for his valuable contributions during fieldwork. I especially want to thank Ruth for her patience and support.

Thanks also go to BTM Consult for granting permission to reproduce tables from their *World Market Update 2005*.

Series Editor Preface
Energy, Climate and the
Environment

Concerns about the potential environmental, social and economic impacts of climate change have led to a major international debate over what could and should be done to reduce emissions of greenhouse gases, which are claimed to be the main cause. There is still a scientific debate over the likely scale of climate change, and the complex interactions between human activities and climate systems, but, in the words of no less than Governor Arnold Schwarzenegger, *"I say the debate is over. We know the science, we see the threat, and the time for action is now."*

Whatever we now do, there will have to be a lot of social and economic adaptation to climate change-preparing for increased flooding and other climate related problems. However, the more fundamental response is to try to reduce or avoid the human activities that are seen as causing climate change. That means, primarily, trying to reduce or eliminate emission of greenhouse gasses from the combustion of fossil fuels in vehicles and power stations. Given that around 80% of the energy used in the world at present comes from these sources , this will be a major technological, economic and political undertaking. It will involve reducing demand for energy (via lifestyle choice changes), producing and using whatever energy we still need more efficiently (getting more from less), and supplying the reduced amount of energy from non-fossil sources (basically switching over to renewables and/or nuclear power).

Each of these options opens up a range of social, economic and environmental issues. Industrial society and modern consumer cultures have been based on the ever-expanding use of fossil fuels, so the changes required will inevitably be challenging. Perhaps equally inevitable are disagreements and conflicts over the merits and demerits of the various options and in relation to strategies and policies for pursuing them. These conflicts and associated debates sometimes concern technical issues, but there are usually also underlying political and ideological commitments and agendas which shape, or at least colour, the ostensibly technical debates. In particular, at times, technical assertions can be used to buttress specific policy frameworks in ways which subsequently prove to be flawed.

The aim of this series is to provide texts which lay out the technical, environmental and political issues relating to the various proposed policies for responding to climate change. The focus is not primarily on the science of climate change, or on the technological detail, although there will be accounts of the state of the art, to aid assessment of the viability of the various options. However, the main focus is the policy conflicts over which strategy to pursue. The series adopts a critical approach and attempts to identify flaws in emerging policies, propositions and assertions. In particular, it seeks to illuminate counterintuitive assessments, conclusions and new perspectives. The aim is not simply to map the debates, but to explore their structure, their underlying assumptions and their limitations. Texts are incisive and authoritative sources of critical analysis and commentary, indicating clearly the divergent views that have emerged and also identifying the shortcomings of these views. However the books do not simply provide an overview, they also offer policy prescriptions.

The present volume looks at a technology which has attracted its fair share of controversy. While enthusiasts look to wind power as one of the main sources of renewable energy, objectors are concerned about the visual impact of wind farms and the costs and operational reliability of systems which rely on intermittent energy inputs. These concerns are no longer of marginal importance. We now have a wind power industry employing around 300,000 people around the world, and over 60 gigawatts of generation capacity – most of it in the European Union. If this industry is to continue to expand, the technical economic and environmental uncertainties have to be resolved. This book provides an overview of the strategic debate about the development of wind power in Europe, looking in detail at the conflicts over financial support mechanisms, interpretations of public acceptability and the issue of intermittency. In challenging some existing beliefs, it attempts to open up the debate over how this new energy option should be developed.

List of Abbreviations

ADEME	Agence de l'environnement et de la maîtrise de l'énergie (the French Environmental and Energy Efficiency Agency)
AEE	Asociación Empresarial Eólica (Spanish Wind Energy Association)
AONB	Area of outstanding natural beauty
APPA	Asociación de Pequeños Productores y Autogeneradores (Spanish Association of Small Producers of Renewable Energy)
ART	Average Reference Tariff (of electricity in Spain)
BETTA	British Electricity Trading and Transmission Arrangements
BLS	Bundesverband Landschaftschutz (German Association for Landscape Protection)
BMU	Bundesministerium für Umwelt, Naturschutz und Reactorsicherheit (German Federal Ministry for the Environment, Nature Conservation and Nuclear Safety)
BNA	Bundesnetzagentur (German energy regulator)
BWE	Bundesverband WindEnergie (German Wind Energy Association)
BWEA	British Wind Energy Association
CCGT	combined cycle gas turbine
CCL	Climate Change Levy
CDU	Christlich Demokratische Union (Christian Democratic Union – German political party)
CECRE	Centro de Control de Régimen Especial (Spanish Control Centre for the Special Regime – part of REE, see below)
CEGB	Central Electricity Generation Board (the former UK nationalised electricity utility)
CENER	Centro Nacional de Energías Renovables (Spanish National Centre for Renewable Energy)
CEO	chief executive officer
CESA	Corporación Eólica SA (Spanish developer now owned by Acciona)
CHP	combined heat and power

CNE	Comisión Nacional de Enérgia (Spanish energy regulator)
CNSE	Comisión Nacional del Sistema Eléctrico (Spanish National Commission for the Electrical System)
CPRE	Campaign for the Protection of Rural England
CPRW	Campaign for the Protection of Rural Wales
CRE	Commission de régulation de l'électricité (French electricity regulator)
CSU	Christlich-Soziale Union (Christian-Social Union – German political party)
DENA	Deutsche Energie-Agentur (German Energy Agency)
DKK	Danish krone (unit of currency)
DM	Deutsch Mark (former unit of currency in Germany)
DTI	Department of Trade and Industry (UK)
DWIA	Danish Wind Industry Association
DWTOA	Danish Wind Turbine Owners' Association
E.ON	German electricity utility
ECJ	European Court of Justice
EdF	Electricité de France (French utility)
EEA	European Environment Agency
EEG	Erneuerbare Energien Gesetz (the German Renewables Energy Sources Act of 2000)
EHN	Energía Hidroelectrica de Navarra (Spanish utility)
EIA	environmental impact assessment
EM	ecological modernisation
EnBW	Energie Baden Württemberg AG (German utility)
ESI	electricity supply industry
EU	European Union
EU-ETS	European Union Emissions Trading Scheme
EUFER	Enel Unión Fenosa Renovables (Spanish utility)
EWEA	European Wind Energy Association
FDP	Freie Demokratische Partei (Free Democratic Party – German political party)
FEE	France Energie Eolienne (French wind energy association)
FoE	Friends of the Earth
GHG	greenhouse gas
GW	gigawatt
GWEC	Global Wind Energy Council
GWh	gigawatt hour
HT	high tension
HV	high voltage

IDAE	Instituto para la diversificación y ahorro de la energía (Spanish Institute for Energy Diversification and Efficiency)
IEA	International Energy Agency
ISET	Institüt für Solare Energieversorgungstechnik (German Solar Energy Institute)
ITER	Instituto Technológico de Energías Renovables (Spanish Technological Institute for Renewable Energy)
kW	kilowatt
kWh	kilowatt hour
MEDD	Ministère de l'écologie et du développement durable (French Ministry for Ecology and Sustainable Development)
MINER	Ministerio de Industria y Energía (Spanish Ministry for Industry and Energy)
m/s	metres per second
MW	megawatt
MWh	megawatt hour
NETA	New Electricity Trading Arrangements
NFFO	non-fossil fuel obligation
NFPA	Non-Fossil Purchasing Agency
NGO	non-governmental organisation
NPPG	National Planning Policy Guideline (in Scotland)
O&M	operations and maintenance
ODPM	Office of the Deputy Prime Minister (UK)
Ofgem	The Office of Gas and Electricity Markets (UK)
OMEL	Operador del mercado eléctrico (Spanish electricity market operator)
OOA	Organisationen til Oplysning om Atomkraft (Danish antinuclear organisation)
OSS	Open Spaces Society
OVE	Organisationen for Vedvarende Energi (Danish Organisation for Renewable Energy)
p.a.	per annum
PER	Plan de energías renovables (Spanish Renewable Energy Plan 2005–2010)
PPA	power purchase agreement
PPG	Planning Policy Guidance (in England and Wales)
PPS	Planning Policy Statement (in England and Wales)
PV	photovoltaic
R&D	research and development
RA	Ramblers' Association

RD	Real Decreto (Spanish royal decree)
REE	Red Eléctrica de España (Spanish national grid company)
REF	Renewable Energy Foundation
REFIT	renewable energy feed-in tariff
RES	renewable energy sources
RES-E	renewable energy sources (generating) electricity
RO	Renewables Obligation (UK)
ROC	Renewables Obligation Certificate
RPS	renewables portfolio standard
RSPB	Royal Society for the Protection of Birds
RWE	German electricity utility
SD	sustainable development
SEO/BirdLife	Sociedad Española de Ornitología (Spanish Ornithological Association)
SER	Syndicat des énergies renouvelables (French renewable energy association)
SPA	Special Protection Area (protected site created under EC Directive 79/409/EEC on the conservation of wild birds)
SPD	Sozialdemokratische Partei Deutschlands (Social Democratic Party – German political party)
SPPEF	Société pour la protection des paysages et l'esthétique de la France (French landscape and heritage NGO)
StrEG	Strom-Einspeisungs-Gesetz (German Electricity Feed-in Law of 1991)
TSO	transmission system operator
TW	terawatt
TWh	terawatt hour
UNESA	Asociación española de la industria eléctrica (Spanish Electricity Industry Association)
VdC	Vent de colère (French antiwind organisation)
VDEW	Verband der Elektrizitätswirthschaft (German Association of the Electricity Supply Industry)
VDMA	Verband Deutscher Maschinen- und Anlagenbau (German Engineering Federation)
VDN	Verband der Netzbetrieber (German Association of Electricity Network Operators)
VIK	Verband der Industriellen Energie- und Kraftwirthschaft (German Association of the Industrial Energy and Power Industry)
VoS	Views of Scotland (Scottish antiwind organisation)
WWF	World Wide Fund (for Nature)

1
Contextualising the Wind Power Debate

Introduction

Wind power has become an emblematic topic in debates on energy sourcing, climate change and sustainable development. The three-bladed 'Danish concept' of wind turbine has come to represent innovative and sustainable energy sourcing. It has become a symbol of hope and an affirmation of belief in a cleaner future. However, wind power's status as a green icon can obscure its material reality. Wind farms have impacts on the countryside, on rural communities, on grid management and configuration, and on the electricity industry. Wind turbines are the largest structures to be found in the rural landscape, and each generation grows larger. A 2.5 MW turbine has a 60 metre diameter rotor placed on a 100 metre tower, producing a total height from ground to blade tip of some 130 metres. In comparison, the major architectural reference point of the village church spire is dwarfed. In addition, the blades of the wind turbine move, and the human eye is attracted to any movement. The steady, symmetrical rotation of turbine blades is unique. The combination of size and movement produces a major visual impact, whilst the infrastructure required to build, cable and grid connect a wind farm has a significant landscape impact. Reactions to these impacts vary enormously. Compare for example Jonathan Porritt's description of wind turbines as 'objects of compelling beauty' with Bernard Ingham's statement that 'steel bog brushes in the sky (...) are not sustainable'.[1]

Differences of view can quickly turn into controversy and lead to polarisation. As Mark Jaccard pointed out, the energy sector is rife with hotly contested opinions:

> People tend to feel passionate when it comes to energy; they often fit into camps that are dead against nuclear or fanatical about

1

renewables or suspicious of energy efficiency or faithful to fossil fuels. My research for this book suggests to me that such passionate prejudgments about our energy options can dangerously distort our view of the evidence (Jaccard, 2005: 28).

Discussions on wind power offer many examples of such partisan tendencies and their associated problems. To avoid throwing more fuel on the fire, the present analysis will adopt as dispassionate and impartial a view as possible.

The purpose of this study is to take stock and encourage informed debate. It aims to analyse the political, economic and social processes which explain the deployment of wind power, and to extrapolate lessons and recommendations for the future. More specific objectives include contributing to social learning in the areas of improved policy design for renewables, revised institutional arrangements and the mapping of routes to societal engagement. The analytical perspectives adopted are drawn mainly from the social sciences and policy studies.

Baumgartner and Midttun (1987b: 291) argued that 'the role of the social sciences (...) is to make people aware of options and alternative paths of development'. This approach informs the core perspective of this book which is to identify and explain the development paths taken by the wind power sector in the European Union, with particular reference to five nations: Denmark, Germany, Spain, France and the United Kingdom. Of these, the first three are acknowledged to be international leaders in the sector, whilst the latter two are late arrivals to it – as are most countries in the world. Comparison between the dynamics of leadership and followership will also be of relevance to a large number of other nations, even though their cases cannot be given detailed treatment in this volume.

Identifying development paths

But what are the development paths and choices in question? Specific paths for renewables will next be identified in theoretical terms, whilst later chapters will use theory to understand national practice and to explore future choices.

Choices can be structured in a variety of manners. Here I introduce three complementary structurations, which will be explained in detail below. The first arises from *socio-economic choices,* with a choice between a 'hard path' or a 'soft path'. The second is in terms of *policy frame choices*. In promoting renewables, the stress sometimes falls *either* on economic issues *or* on environmental issues. But the economic and environmental policy frames can also be integrated in various manners. Their integration produces a third structuration in terms of *governance choices*. Theories of 'sustainable development' and 'ecological modernisation' each integrate policy frames into a particular governance perspective, but their orientations and outcomes differ markedly. Consequently I argue that they present alternative development trajectories. Applying these three structurations of choice to wind power will allow clearer understanding of development paths taken to date and the options for the future.

Socio-economic choices: 'hard' or 'soft' paths?

In the energy domain, the work of Amory Lovins articulated a model of technology choice based on the alternative between a 'hard path' and a 'soft path':

> The first path (…) relies on rapid expansion of centralized high technologies to increase supplies of energy, especially in the form of electricity. The second path combines a prompt and serious commitment to efficient use of energy, rapid development of renewable energy sources matched in scale and in energy quality to end use needs, and special transitional fossil fuel technologies. (Lovins, 1977: 24)

Lovins (1977: 38–9) specified the five defining characteristics of 'soft technologies' as:

1. reliance on renewable energy flows, rather than depletable energy capital;
2. diversification by the aggregation of large numbers of small contributions;
3. flexibility based on (relatively) 'low tech', which was 'accessible rather than arcane';
4. a match between scale of sourcing and geographic distribution with end-use needs;
5. a match in energy quality to end-use needs e.g. thermal energy where heat is required (rather than electricity generation).

In his view, 'soft technologies' have the merit of being flexible, dispersed and locally tailored. They have short build times, being based on modular construction. They distribute technical risks across a range of technologies whilst minimising economic and social risks, including those arising from error, accidents or sabotage (Lovins, 1977: 50–1). For these reasons they are more forgiving than large thermal installations – especially nuclear plants, which for Lovins constituted the epitome of the hard path. Additional problems of the 'hard path' included interventionist and bureaucratic central control and the concentration of political and economic power in the hands of major, oligopolistic companies backed by large government agencies, with a consequent undermining of market mechanisms (Lovins, 1977: 148–51). To help clarify the differences between these paths, the contrasting features are summarised in Table 1.1. Like many opponents of nuclear power in the 1970s, Lovins believed that hard paths led to the erosion of democracy and an 'industrial-military complex'. Thus he argued that technology choice necessarily entailed a selection between societal development pathways.

Lovin's analysis was prescient and remains relevant, although not all of his predictions have been confirmed. Accelerating depletion of energy resources and growing environmental problems are a reality. But the expectation that nuclear power would lead to authoritarian regimes has not been fulfilled in Western nations, whilst the domination of energy sourcing by corporate oligopolies is now extending from conventional to renewable sources.

The vision proposed by Lovins arose within the 'alternative technology' movement of the 1970s. As argued by Dickson (1974: 11): 'this technology would embrace the tools and machines necessary to reflect and maintain non-oppressive and non-manipulative modes of social production, and a non-exploitative relationship with the natural envi-

Table 1.1 Comparing 'hard' and 'soft' paths

'hard paths'	'soft paths'
depletable energy capital	renewable energy flows
centralised and interventionist	decentralised and market oriented
inflexible, one-off, large scale	flexible, modular, locally tailored
large utility	local ownership
marginalisation and alienation of citizens	participation, 'conviviality'
high risk	low risk

ronment'. Opting for alternative technologies involved more than changes in techniques or tools. It meant taking ecological responsibility and making large-scale reforms in social and political organisation. However, the radical economic and social ambitions of the 'alternative technology' movement have not been fulfilled. Nevertheless, its optimistic, reformist views have continued to colour understandings of renewables. An example is Susan Baker's contention that:

> the ecocentric approach espouses 'appropriate' technology; that is technology that is in keeping with natural laws, small in scale, understandable to lay people and workable and maintainable by local resources and labour. This is also closely connected with a belief in community empowerment achieved through generations of community or 'grassroots' consciousness, and improvement in environmental quality through co-operative endeavours and local initiatives. It is symbolised by the Danish community-owned wind farms. (Baker *et al.*, 1997: 11)

This vision of sustainable development is close to what Bell (2004: 98) called 'the "green" ideal of small, self-sufficient agrarian communities'. This constitutes the first development option for the wind sector. For many activists in Denmark and Germany, wind power is associated with small-scale installations owned by farmers, or rented out by them on agricultural land. Further, this model offers the prospect of 'embedded generation' (e.g. linked directly into the distribution grid) with electricity generated, consumed and managed locally. Although community-led developments in Denmark and Germany are far from constituting a reform of market capitalism, vestiges of the radical ideology of the 'alternative technology' movement still informs the discourse of (some) wind power enthusiasts. But the small-scale, communitarian approach is not the only mode of wind power implementation.

The mainstream model of electricity provision relies on 'bulk power'. The 'bulk power' option stresses a culture of electricity production and consumption based on mass industrialisation, and has much in common with a 'hard' path. Its key components are (a) large-scale generation leading to (b) low costs. The purpose of cheap 'bulk power' is to facilitate mass production and low prices across the whole economy in a 'virtuous spiral'. It is based on centralised thermal plants (fossil and nuclear) located in or near big cities to service major industrial and population concentrations, with a 'hub and spokes' grid configuration as a by-product. This option has resulted in a production-oriented culture of output maximisation and of driving down costs in a bid for 'efficiency' – although 'efficiency' is understood very narrowly, given that conventional power plants are dedicated to the generation of electricity and simply throw away the heat they produce. Centralised 'bulk power' has been based on particular institutional frameworks, typically involving (until the recent past) government intervention and top-down planning, often through the medium of nationalised companies. This combination of elements produced a 'winning formula' for many industrialised nations in the twentieth century. It became part of the mind-set of economic and political elites, governing their decisions on technology choice. Moreover, because electricity has become a necessity, 'keeping the lights on' is a political imperative.

A second development option for the wind sector is incorporation into this 'bulk power' model, by the same actors and within the same institutional frameworks as the conventional generation technologies that constitute the 'hard path'. In the UK, the use of wind energy for centralised electricity generation was considered relatively early by the CEGB, as the study by Taylor (1983: 45–7) showed. However, this option did not become commercially viable till the 1990s. But by the 2000s, large-scale wind farms – both onshore and offshore – are increasingly being developed by electricity utilities as a variety of 'bulk power'. This constitutes an attempt to maintain the prevailing model of consumption, whilst changing only the model of production and transmission. Thus the 'bulk power' model implemented by large firms and international utilities contrasts with the 'alternative technology' model based on small-scale, local ownership. The former is a variant of the 'hard path', whilst the latter represents a continuation of the 'soft path'. For the future, an important question is whether both development paths will continue to unfold, or whether one is superseding the other.

Policy frame choices: prioritising the economy or the environment?

Technology choices can be motivated by different reasons. In a 'soft path' perspective, priority is usually given to environmental and social factors, whilst in a 'hard path' perspective the stress is on the economy. This difference in perspectives has produced two key policy frames underpinning the development of renewables in the recent period: the environmental frame and the economic frame. A 'policy frame' identifies an issue which is capable of resolution through political and administrative means. Majone (1989: 23–4) contended that 'the most important function both of public deliberation and of policy-making is defining the norms that determine when certain conditions are to be regarded as policy problems'. An issue identified as a policy problem may be addressed in a variety of manners. At the cognitive level, framing is a process of 'selecting, organising, interpreting, and making sense of a complex reality to provide guideposts for knowing, analysing, persuading and acting' (Rein and Schön, 1993: 146). Policy framing is therefore the process by which choices of means and ends are articulated, and involves a combination of cognitive and empirical approaches to finding policy solutions.

The economic policy frame

Since the industrial revolution, humanity has seen massive economic development and growth. Bulk energy has been the driver of growth. The major sources of bulk energy have been fossil fuels. Because of the stress on economic growth, ensuring access to bulk energy sources continues to be a core aim not just of economic and industrial policy-making, but of international geopolitical strategies. In the twentieth century, the energy policy of industrialised countries was typically framed as a question of supply. The aim was the rapid exploitation of coal mines, oil wells and gas fields at home and, where this was inadequate, in energy-rich regions such as the Middle East, North Africa and Russia. International energy sourcing raised issues of security of supply, encompassing not only the physical questions of access, storage and transportation, but also the geopolitical dimensions of market regulation, actor strategies and price formation. However, during the middle of the twentieth century massive imports of oil created a relationship of dependence between industrialised nations and oil-rich countries. The oil crises of the 1970s and 1980s provoked an acute awareness of the risks of high reliance on imported energy. It reframed the security of supply concept as an aspiration to greater national 'self-sufficiency' or 'independence' in energy sourcing. The ideology of 'national independence' was particularly marked in France and was used to justify expansion in nuclear power. Elsewhere, opportunities for increased domestic production were seized on, for example by the UK and Denmark in offshore oil and gas. Germany and Spain propped up ailing coal industries with state subsidies for security of supply and employment reasons, but lacked domestic sources of other fossil fuels.

These contextual elements created unstable framework conditions for use of renewable energy sources (RES), with problems over the sourcing of conventional energy tending to encourage their development whilst periods of fossil fuel abundance tending to discourage it. Further, they led to 'taken-for-granted' assumptions regarding energy use. The most energy-intensive sources were preferred, namely fossil fuel and nuclear. Low energy intensity sources – such as renewables – were considered either as simply inadequate for the task of driving industry and the economy, or else as ancillary and minor sources. However, as Lovins argued, energy consumption raises the question of appropriateness, namely the match between sourcing and usage. For example, wood and gas are appropriate sources of space heating, whereas electricity for heating is a highly inefficient use of energy – given that conventional generation involves discarding large quantities of heat. In the late twentieth century the combination of abundant sourcing and low prices meant that the (mis)match between energy sources and energy uses was only taken seriously at times of shortage and quickly forgotten in times of glut. Efforts to improve energy efficiency have followed the same cycle. In consequence, efforts to expand renewables and improve energy efficiency have been extremely limited.

But can abundance and wasteful consumption continue forever? In their famous *Limits to Growth* thesis, Meadows *et al.* (1972) predicted rapid exhaustion of fossil fuels. Their predictions proved inaccurate, but the long-term problem they posed has not gone away. In the recent period, analysts have preferred to talk in terms of oil and gas 'peaks', followed by progressive depletion. Uncertainties persist over when this will happen, but major price fluctuations can be expected as competition for energy sources increases. Over 2005–6, oil prices hit record highs of around $70 a barrel due to geopolitical tensions in producer regions, notably Iraq and Iran, and increased demand from industrialising nations, especially China and India. For the time being, this has not resulted in recessions comparable to those triggered by the oil price shocks of the 1970s. Changes in technology and economic structures have cushioned the impact. But for the long term the consequences of depletion are clear: by the late twentieth-first century it is unlikely that fossil fuels can drive worldwide economic expansion on the traditional model. This will leave governments with three options: (1) monopolise conventional energy resources by diplomacy or war, (2) find alternative sources of bulk power, or (3) change the model of economic and social development.

In the 2000s, the first two options have found favour, but the third has not. Alongside the attempt to secure their access to oil and gas in distant parts of the world, industrialised nations are seeking to diversify their energy sourcing and reduce dependence on imports by recourse to renewables. Renewables are an indigenous and inexhaustible energy source. They can provide fuel for heating and transport, as well as generate electricity. Consequently, their potential is being reassessed. Many now consider that their time has come. However RES technologies remain underdeveloped. During the twentieth century, low conventional energy prices generally made it uneconomic to invest in technological innovations related to RES sourcing, conversion and consumption. Despite the obstacles, some RES technologies have been developed through pioneering work by enthusiasts and supported by public subsidies. However, their fitness for use varies across a spectrum from 'emergent' to 'near market' to 'mature'. Large-scale hydroelectric power is an example of a mature RES technology. Marine technologies remain 'emergent'. Wind power in the 1980s and 1990s was in the 'near market' stage, but in the 2000s made rapid strides towards 'maturity'. In addition, its rapid deployment has given wind power a 'trailblazer' role for other renewables due to come on stream in the next decade. Thus the renewables sector is sometimes conceived as a new component of the economic frame for energy policy-making, supporting and extending conventional sources of bulk power and helping to drive economic growth.

The environmental policy frame

A 'traditional' environmental frame concentrated on the mitigation of harm to the environment, habitats and wild life caused by human activities. In relation to conventional energy sources, environmental damage can arise at every phase of the operational life-cycle. The extraction phase causes landscape degradation and water pollution. Drilling and mining involve industrial accidents and fatalities. In the combustion phase, fossil fuels are responsible for air pollution problems including 'acid rain', with harm to buildings, forests and ecosystems, and also to human health. Moreover, the release of greenhouse gases (GHG) from fossil fuels is a cause of climate change. Conventional energy sources also pose problems regarding the disposal of unwanted by-products. In particular, the long-term storage of nuclear waste has yet to find a definitive solution, whilst the decommissioning of nuclear reactors poses unresolved issues. To whatever extent these problems are bequeathed to future generations, it imposes costs on them. These adverse impacts are variously termed 'environmental externalities' or 'social costs'. Because they are largely unpaid, the true costs of fossil fuels are disguised and are significantly higher than their market prices. However, analyses attempting to quantify externalities in monetary terms have produced large variations in estimates.[2]

Over and against the external costs created by conventional energy sources, renewables are frequently described as 'safe' and 'clean'. This is an important argument in their favour. But this is not to say that externalities never arise, only that they are of different varieties and intensities. As noted by Lee (2002: 107) 'there is no such thing as a "clean" energy source with respect to the environment, but some energy conversion technologies are friendlier to the environment than others'. One illustration is large-scale hydroelectric power, where the building of dams and the flooding of valleys have serious consequences in terms of ecological impact and population displacement. Wind power too is accused of depredation to the landscape, harm to wild life and adverse effects on rural communities. Although the negative environmental and social impacts of wind power have been significantly less serious than those arising from conventional energy technologies, they cannot be discounted to zero. On the other hand, the environmental benefits of wind power are clear: no extraction costs, no delivery implications in terms of transportation and logistics, and an absence of GHG and other emissions at the point of generation. In the context of the 1997 Kyoto Protocol, by which industrialised nations committed themselves to reductions in GHG emissions, recourse to renewables has been presented as a means to achieve targets. Thus the environmental policy frame seeks to address and prioritise a range of energy issues running from sourcing to consumption.

Governance choices: 'sustainable development' or 'ecological modernisation'?

Policy debates in the 1970s often posited a *conflict* between economic and environmental frames, with the assumption that one must be sacrificed to the other. However, the 'sustainable development' (SD) and the 'ecological modernisation' (EM) concepts have posited compatibility between economic growth and environmental protection, albeit in significantly different ways.

The SD agenda systematically links environmental problems and development issues. Although SD is acknowledged to be an open and contested concept, a number of elements have common currency.[3] The 1987 Brundtland report offered the now canonical definition of the new approach as 'development that meets the needs of the present without compromising the ability of future generations to meet their own needs' (World Commission on Environment and Development, 1987: 43). The SD concept rests on 'three pillars' – economic growth, environmental protection and social development – which are conceived as complementary and non-contradictory goals. Social development involves the reduction of poverty, social exclusion and injustice, and the promotion of community involvement in decision-making. Environmental protection is understood ambitiously as ecological renewability with respect to resource management, energy sourcing and climate policy. Economic growth, however, is often assumed by policy-makers to mean the traditional concept of GDP increases through free trade in the market, though this interpretation is contested by social movement activists. Due to these features, SD is both a continuation of well-rehearsed debates and an ambitious new departure.

The integration of economic, environmental *and* social concerns into policy-making is a hall-mark of the SD approach. It opened the door to greater recourse to deliberative and inclusionary processes, involving a broader cross-section of the population. Calls to develop this societal dimension were taken up at the 1992 Rio world conference, integrated into the Rio Declaration (principles 10 and 20 to 22) and developed subsequently. In the EU, the sustainable development framework has led to a 'horizontal' conceptualisation of policy-making to complement the sectoral or 'vertical' perspective which has been the norm (Aguilar Fernández, 2003). Key EU initiatives include the Sixth Environmental Action Programme – 'Environment 2010: Our Future, Our Choice' – which called for a deepening of environmental policy integration across a range of sectors, such as agriculture, industry and energy.

On the other hand, ecological modernisation (EM) offers a more narrowly industrial conceptualisation of the integration of environmental issues into the economy. It argues that economic growth is fully compatible with environmental protection, and moreover that the two are mutually reinforcing. Thus improved environmental performance is considered to lead to greater economic competitiveness and industrial innovation; this improves environmental performance and continues the virtuous spiral. More specifically, it identify pathways to the internalisation of environmental costs through greater 'factor productivity' and the 'decoupling' of energy input from growth.[4] EM has moved from being a sociological theory explaining recent developments in industrial society to a political programme with its accompanying raft of public policies (Spaargaren and Mol, 1992: 334). In practice, the EM concept has given support to 'business-as-usual' models stressing economic growth, whilst addressing environmental problems caused by heavy industry in nations such as Germany and the Netherlands, the two countries in which this theory was developed and has been most influential. Thus Hajer (1995: 32) observed that EM is 'basically a modernist and technocratic approach to the environment that suggests that there is a techno-institutional fix for the present problems'. Because the practice of EM has displayed a strong market orientation, Johnson (2004) argued that EM fitted well with neo-liberal globalisation driven by free trade and multinational corporations. In combination, these elements add up to a programme of industrial renewal based on ecological restructuring that can be termed an 'ecomodernist growth paradigm'. On the other hand, EM does not deal with societal issues such as community participation, environmental injustice or intergenerational equity, as does sustainable development. Indeed, it is debatable whether this form of modernisation constitutes *development* in the social and political senses. In consequence, Langhelle (2000: 303) argued that 'ecological modernisation should, therefore, be seen as a necessary but not sufficient, strategy for sustainable development, and the two concepts should not be conflated'.

In summary, the 'sustainable development' and 'ecological modernisation' theories both seek synergy between the key policy frames of the economy and the environment but integrate them differently. SD places stress on social development, but this dimension is largely omitted from EM. A different category of actors is privileged by each concept, with the stress falling on social actors in SD and on economic actors in EM. Hence the two concepts offer distinct ways of understanding the development and role of renewable energy. A case can be for made for wind power as an example of ecological modernisation,[5] or as a pioneering example of 'sustainable development'. Later chapters will therefore assess which of the two theories gives a more accurate portrayal of the evolution of the wind power sector.

Wind power and development path choices

Three sets of development path choices have been identified: socio-economic choices, policy choices and governance choices. If we represent these sets of choices as upward steps on a ladder of increasing decisional complexity, the question arises of how we move from one rung to the next, and how choices taken in relation to a lower rung impact on higher level outcomes. In theory, this universe of choice could be mapped in a systematic manner, inter-relationships identified, consequences predicted and decisions taken on the basis of a hierarchy of priorities. But in the real world, decision-making is rarely so self-reflexive, deliberate or explicit. Indeed, even the decision-making forum and the identity of the decision-makers can be elusive. Everyday decision-making is generally circumscribed and incrementalist, with outcomes resulting from the combination of many small adjustments made by independent actors, rather than reflecting a coordinated 'master plan'. Choices made in these conditions are characterised by 'bounded rationality', and have unpredicted and unpredictable consequences. Analytical frameworks have therefore to respond to this 'organised chaos'. Accordingly in the next section I consider some of the pragmatic dimensions of evolutionary and open-ending decision-making. To do so, I will first contextualise wind power within the operational frameworks of the electricity supply industry. An analysis based on the concept of 'path dependence' will trace how development trajectories evolved in the past, in order to understand how historical choices impact on future decisions.

From social contract to social acceptability

The electricity sector and the social contract concept

Haugland, Bergesen and Roland (1998: 16) argued that the energy sector can be analysed on the basis of three interlocking approaches, namely (1) the energy and environmental policies which provide framework conditions; (2) the structure of energy industries which influence the ways in which actors respond to new challenges, including policy formulation and implementation, and (3) the varieties of social contract which provide institutional and behavioural rules. In later chapters, we will return to questions of policy, institutions and industry structure. For the moment, I focus on the concept of social contract which they defined in the following way:

> there must be a fundamental common understanding between the energy industries and society at large, represented by governments. This is the crux of the social contract that binds together the policy and industrial levels described above, by defining the rights and obligations of both sides, Through this 'quid pro quo', actors in the sector are given societal objectives to which they are committed – security of supply, employment, environmental goals, etc. In return, they acquire a carefully defined freedom of action. This can take the

form of legal or quasi-legal monopoly of rights, for example within a specific area. Governments can favour one energy source at the expense of another, through taxation in return for regional development, employment, or some other laudable goal. An energy company can in addition be given access to the treasury by way of state aid or hidden subsidies (Haugland, Bergesen and Roland, 1998: 18–19).

Such social contracts have been in place in the electricity sector for many decades, but have varied in form and content over time and within national contexts. Some components have been formally or legally defined. For example, the notion of electricity provision as a 'public service' is highly developed in the French case, with a historical requirement for prices to households to be the same throughout France. But other components simply constitute custom and practice, arising out of particular technological, industrial and political contexts. A core example was a tendency in European countries in the twentieth century to allocate territorial monopolies (sometimes national, sometimes regional) to electricity utilities and, in many cases, for utilities to be in public ownership. This pattern of industry ownership and structure led to the development of particular expectations, values and norms. The public service ethos has already been mentioned. Expectations arouse regarding the nature and siting of electricity generation plants and associated grid infrastructure. Large-scale power plants were constructed close to 'load centres', namely concentrations of business and household consumption. These were usually towns and cities, producing a 'hub and spokes' design with grid infrastructure radiating out from urban to rural locations. This pattern of development was a major norm within the sector. Settlements in peripheral areas were the last to be connected to mains electricity, usually on the basis of low-voltage cabling. This resulted in the phenomenon of 'weak grids' in many rural areas, whilst the interconnection of major load centres required 'strong grids' using high-voltage power lines. In brief, associated with the rise of the electricity sector were norms based on industrial context (large-scale generation plants), governance context (often characterised by public ownership) and a public service ethos (with security of supply, obligatory provision and equitable price formation as guiding principles). This set of norms constituted a form of social contract.

Yet a social contract in relation to electricity were never drawn up as a legal document. Norms evolved *ad hoc* and their application changed over time. Major changes in the social contract were initiated by the liberalisation processes of the 1980s and 1990s. Liberalisation of energy markets, including electricity, led to deregulation and privatisation of utilities. The process changed industry structures and conduct. The old monopolies were progressively dismantled and suppliers were forced to compete. Opening to market competition gave consumers greater choice over supplier. But it also reopened issues that had been settled under the previous social contract pertaining to a context of monopoly provision. For example, how were societal desiderata on matters such as uniform pricing for households and protection from disconnection for the economically disadvantaged to be settled in the era of liberalisation? A regulator is essential to ensure that suppliers do not cream off the best accounts and disconnect the least profitable. Re-regulation, through deepened institutionalisation of the regulation function (as undertaken by bodies such as Ofgem, CRE, BNA and so forth), has been the means to tackle these issues. It has contributed towards the formation of new 'social contracts', arguably on a more legally constituted basis.

The evolution of the electricity sector has been affected by other contextual, technological and political developments. These have impacted on norms and expectations within the 'social contract' in less clear-cut manners than in the case of liberalisation. The contribution of fossil fuel combustion to climate change has raised difficult questions over energy sourcing. In the electricity sector, this has led to vexed debates. Civil society actors are seeking to redefine norms related to sourcing. NGOs such as Friends of Earth and Greenpeace are increasingly stigmatising coal-fired generation, whilst continuing to oppose nuclear power. In relation to technology, technical progress has meant that medium-scale generating plants have become economically competitive with large-scale units. This development presents new opportunities in terms of facility siting and grid configuration, but also raises new challenges. Facility siting is regulated by institutional norms embodied in land-use planning processes. Indeed, the planning regime itself constitutes a variety of social contract. However, in relation to the electricity supply industry, the norms which planning currently embodies evolved in the technological, social and political contexts of the twentieth century. Those norms have been disturbed by technological innovations, new ecological imperatives and changing societal preferences. Responses to disturbances of old norms and the establishment of new ones are being worked through, on an *ad hoc* basis.

A set of norms creates boundary conditions for what is considered acceptable or unacceptable. Over the course of the twentieth century, it became increasingly unacceptable to households in rural areas around Europe to be without mains electricity. But electrification also raised problems. There has been a long history of resistance by rural dwellers and amenity organisations to high-voltage power lines which interconnect load centres, especially where power lines traversed nature conservation areas and scenic landscapes.[6] Large pylons in the countryside were interpreted by some parties as a transgression of norms related to the electricity sector. This justified their resistance. In bringing norms and transgression of norms into the frame, a social contract also introduces conditionality. In other words, decisions and actions are considered acceptable on the condition that they respect behavioural norms. This can be illustrated with two examples. Access to mains electricity is a normative condition: failure to provide such provision can be appealed. Fair and consistent pricing is a normative condition: predatory pricing is rejected. These are some of the practical entailments of the abstract concept of 'social contract'. Particular proposals – for construction, transmission, connection, disconnection and so forth – likewise are deemed acceptable or unacceptable on the basis of normative conditions. In summary, the 'social contract' notion contains within it various criteria of conditionality. In the discussion which follows, the term 'social acceptability' will be used to refer to such criteria and conditions.

Wind power offers numerous instances of the problems posed by changing norms. It is a generation technology which – with limited exceptions – has been deployed across the countryside, unlike conventional generation which takes place mostly near or in cities. It brings a new form of economic activity to areas hitherto untouched by industrialisation. The technology has impacts on landscape and on local ecology. It affects the lived experience of rural communities, who are accustomed to the countryside being used for farming but not for electricity generation. These new impacts disrupt long-standing expectations on the proper management of the countryside. They also disrupt the norms of land use planning. In the recent period, many planning authorities in rural locations around Europe have been faced with a wind farm application for the first time. Adapting existing norms to a radically new technology has proved problematic, due to divergent interpretations of the planning 'rule book' arising in contexts of social conflict. In addition, wind power disrupts electricity industry norms, notably with regard to grid configuration and management. Wind farms are often sited in peripheral locations and connected to 'weak grids'. This results in grid saturation and the need to expand grid capacity. The construction of new high-voltage power lines, sometimes crossing ecologically sensitive areas, is once again provoking resistance. Moreover, it is possible that in the next decade the scale of new generating capacity from wind and marine sources built in remote onshore and offshore locations will lead to not just grid reinforcement, but grid reconfiguration. In brief, wind power is displaying a capacity to disrupt social and industrial norms to a significant extent. It may even be in the process of rewriting the social contract related to electricity. Inevitably, this has raised the question of the acceptability of wind power, in different locations, at different scales of deployment and for different reasons.

Exploring the concept of social acceptability

The question of *acceptability* is not the same as the question of *acceptance*. In the context of this book, the issue of acceptability is not addressed to create the presumption that wind power *per se* is unacceptable. Rather, the question concerns the criteria and conditions under which a social, economic or institutional actor decides to accept or reject an idea, vision, proposal or practice. The intent is to understand the conditions under which wind power proposals and patterns of deployment are judged by particular actors. In later chapters those conditions will be explored in detail. But by way of outline, the conditions of acceptability range from practical considerations (such as availability of suitable sites with sufficient wind speeds), to economic issues (such as investment costs, generation prices), to institutional regulation (goals of policy-making, criteria for planning consents), to social and ecological impacts. Assessments based on these empirical conditions come to form an evolving social contract related to wind power. Through the incremental process of social contract formation, choices on wind sector pathways, policies and styles of governance are being made.

Acceptability is not a static but an evolving decision frame. Actors deliberately seek to influence and change the boundary conditions or 'rules of the game'. For example, actors promoting wind power have called for more accommodating planning regimes within which favourable verdicts are given to wind farm applications more predictably, more frequently and above all more quickly. Meanwhile, actors resistant to wind power have called, among other things, for tighter and more formalised strictures on the siting of wind turbines. Where different actors pull in opposing directions a form of intense political competition is created whose object is to *reframe* decision-making norms by *re-negotiation* of social acceptability (i.e. to redefine the conditions or rules under which particular practices or outcomes are deemed to be acceptable or unacceptable).

Why is review of social acceptability important? Firstly, it highlights the fact that acceptance or rejection of new technologies does not arise from subjective whim, but is governed by norms relating to national contexts, traditions and conventions, and to time periods. Secondly, it allows recognition of the distinction between the *categoric* viewpoint of partisan activists and the *conditional* viewpoint of non-partisan participants. In holding a categoric view, partisan activists will simply accept or reject an object of discussion. For example, antinuclear protestors categorically reject nuclear power – in all its forms, in all places, under all circumstances. In contrast, a conditional viewpoint is a judgement arising from the application of an assessment framework to particular circumstances. Because the judgement arises on the basis of evaluative criteria, it may go in favour with one proposal, go against another, whilst requesting improvements from a third. Appreciating the difference between categoric and conditional viewpoints, and knowing what the acceptability criteria are, allows better purchase on slippery debates. Thirdly, in analysing the competition for redefinition of social acceptability, the aim is to *identify* and *make explicit* processes which, being normally conducted at an implicit or indeed 'back office' level, are not fully recognised for what they are, namely a process of social contract negotiation between parties having unequal access to expertise and resources. Once made explicit, characterisation of those processes becomes possible. Fourthly, on these bases, analysis can direct attention to the question of the *legitimacy* of particular types of behaviour that seek to change acceptability – although it may not always be able to resolve that question. This is a inherent difficulty arising where norms and values are themselves in flux. With regard to wind power, precisely because the conditions of acceptability have changed and continue to change, it is not considered possible to settle the debate once and for all. Nevertheless, analysis of the dimensions of acceptability illuminates the debate on the social contract, its associated practices and the norms regulating the behaviour of the various parties. Exploration of these issues allows clarification of *how* development path choices are being taken, by *whom* and *where* they are heading.

Research outline and chapter review

The research process

The analyses presented in this book arose from a lengthy research process. An extensive literature review was undertaken, comprising analysis of published materials from academic and government sources, as well as the 'grey literature' produced by various organisations. Interviews were conducted during 2003–6 in Denmark, France, Germany, Spain and the UK with representatives from the main categories of actor (politicians, policy-makers, energy agencies, wind power companies, trade associations, planners, NGOs, antiwind protesters). Fieldwork also included site visits to a number of wind farms either built, in construction or in planning. Early versions of the present analyses were 'trialled' in a series of conference papers given to specialist audiences in 2004–6, in three articles in academic journals,[7] and in a research report.[8] This process generated peer review and feed-back, which allowed improvements to the methodological and analytical frames. The revised material presented in this monograph was written largely during 2005–6, whilst final editing took place in early 2007.

Chapter themes and research questions

Chapter 2 analyses the major features of the wind sector and sets the scene for the rest of the book. Firstly, it analyses the supply side, identifying critical factors in the international expansion of the wind power industry. Secondly, it considers the demand side by analysing patterns of investment and ownership. Thirdly, it investigates the ways in which supply and demand have co-evolved, since technological innovation and 'upscaling' have changed models of ownership. The chapter thereby addresses the question of how far development paths in the sector have been marked by community ownership and 'alternative technology' ideals or by the 'bulk power' orientation of the electricity supply industry (ESI). Following chapters then provide explanations for development paths, each adopting a particular perspective.

Chapter 3 focuses on mobilisation in favour of wind power. It identifies the actors who have promoted the uptake of wind power, reviews their arguments and strategies and explores the sources of their influence. The underlying questions addressed are: How have wind lobbyists promoted wind power? How have they sought to mobilise policy-makers and public? How have they negotiated questions of social acceptability? What is the extent of their influence in framing policy questions and promoting particular development paths? An underlying question is whether lobbying for wind power forms part of what Elliott (2003: 185) termed a 'hegemonic battle' for market share between alternative energy providers and entrenched fossil fuel interests.

Chapter 4 analyses the content and conduct of policy-making at the national level. It reviews the main policy options available, and analyses the evolution of policy to promote wind power in Denmark,

Germany, Spain, France and the UK, assessing outcomes in each case. It reports critically on the extent to which policy-makers have reviewed and improved national policy on the basis of experience over the long term. Core questions addressed are: What has been the nature and rationale of policy to support renewables, notably wind power? How far has policy been interventionist or 'market-centred'? What are the development path consequences of particular choices of policy design?

Chapter 5 presents cross-national comparisons to draw broader policy lessons. It compares the effectiveness and efficiency of national policy designs in practice, setting them in the context of EU target-setting and European Commission ambitions to achieve 'harmonisation'. The chapter addresses the following core questions: Are policies on track to meet targets? What are the main lessons to be learnt from policy experiments across Europe? What recommendations can be made for improved policy design? Can a single policy template be applied across a number of nations?

Chapter 6 looks at the interaction between wind power and the ESI at several levels. Because wind power is not a stand-alone generation option, it has to be integrated into the ESI at the technical level by grid integration, and at the economic level by market integration. This chapter analyses the consequences in these areas. A core question is the extent to which different electricity industry structures and forms of market organisation impact on the performance and development path of wind power. But wind power deployment also raises issues related to the development trajectories of national ESIs. Will their future direction be decided by market competition or by political prioritisation of energy sources? Both economic and environmental performance are important. Hence this chapter also contextualises the GHG emissions reduction pathways afforded by wind power within the structures of national ESIs. The unifying question in the chapter is this: how do national ESIs and the wind sector inter-act and reciprocally influence the development paths of each other?

Chapter 7 considers the development of wind power 'on the ground'. After reviewing the nature of siting issues, it considers the extent and significance of regional concentration of wind power in the five nations. It goes on to analyse the institutional frameworks for conflict management embedded within national planning traditions. It sets out the issues related to societal participation in planning processes, and the factors influencing social acceptance. The main questions here are: Have siting questions and conflict resolution been handled in a 'bottom-up or 'top-down' manner? What national differences are there in the handling of problem situations, and what can be learnt from cross-national comparison?

Chapter 8 investigates the causes of contestation of wind power, identifying key actors, their preferences, rationale and action repertoire. The chapter's first section deals with 'antiwind' groups whose reason for existence is to oppose wind power. The second section investigates the reactions of long-established organisations – such as nature conservation and amenity associations – who have objected to wind power proposals and analyses their contingent reasons for doing so. By comparing and contrasting the two groups, the third section addresses the question of whether they share a common agenda or have different goals.

The last chapter synthesises earlier discussions. It summarises key findings and policy recommendations related to wind power. The development path taken by the sector is reviewed and characterised, leading to investigation of the political dimensions of renewables and energy sourcing more generally. The contours of future 'social contracts' in energy sourcing are mapped and explored, and their implications for sustainable development are reviewed.

2
Diagnosing the Wind Sector

Introduction

In the space of three decades, the wind sector has grown from its provincial and agricultural origins to a global, high-tech industry. This chapter reviews key aspects of this development, whilst focusing on the wind sector's current characteristics. It thereby provides the context for the thematic analyses of later chapters. The first section provides data on wind power as a global industry. It reviews the turbine industry, identifying the major players and their strategies. The second section turns to the demand side. It considers changing patterns of ownership of wind installations, reviewing the relative importance of small versus large-scale capitalism in the evolution of the sector. The third section analyses the ways in which technological change – particularly increases in size of turbines and installations – has changed both the supply and demand sides of the equation, leading to the transformation of deployment and ownership patterns within the wind sector. The chapter will thereby address the question of whether the development of wind power has been closer to 'the soft path' brokered by the 'alternative technology' movement, or to the 'hard path' and bulk power orientation of the utilities.

Wind power as a global industry

Growth in wind power capacity has spectacularly exceeded expectations. At the start of the 1990s, Sesto and Lipman (1992: 46) offered a projection of 25,000 MW for Europe in 2010. In 1997, the EC white paper on renewable energy set the 'ambitious' target of 40,000 MW by 2010 (European Commission, 1997: 40). Yet by the end of 2005, installed capacity in EU-25 was already 40,504 MW (EWEA, 2006: 6). Although initially limited to Europe and the USA, wind power is now a worldwide phenomenon as Table 2.1 indicates.

However, outcomes in 2005 were exceptional in that the USA was the country to install most new capacity. This was due to a renewal of subsidies in a turbulent context of stop-go policy support. In previous years, Germany followed by Spain have typically seen the largest capacity increases, as Table 2.2 on cumulative capacity indicates.

Table 2.1   New wind power capacity in world markets in 2005

	MW	%
USA	2,431	21.1
Germany	1,808	15.7
Spain	1,764	15.3
India	1,430	12.4
Portugal	500	4.3
China	498	4.3
Italy	452	3.9
UK	446	3.9
France	367	3.2
Australia	328	2.8
Rest of the world	1,507	13.1
World total	11,531	100

Source: GWEC (2006: 11)

Table 2.2   Cumulative wind power capacity in the ten largest markets

Country	2003	2004	2005	Share %
Germany	14,612	16,649	18,445	31.1
Spain	6,420	8,263	10,027	16.9
USA	6,361	6,750	9,181	15.5
India	2,125	3,000	4,253	7.2
Denmark	3,076	3,083	3,087	5.2
Italy	922	1,261	1,713	2.9
UK	759	889	1,336	2.3
China	571	769	1,264	2.1
Netherlands	938	1,081	1,221	2.1
Japan	761	991	1,159	2.0
Total	36,545	42,735	51,686	
percent of world	90.7%	89.2%	87.2%	

Source: BTM Consult (2006: 15)

Table 2.3   **Capacity growth of wind power in the five reference countries**

	Installed MW in 2004	Accumulated MW 2004	Installed MW in 2005	Accumulated MW 2005
Germany	2,054	16,649	1,808	18,445
Spain	2,064	8,263	1,764	10,027
Denmark	7	3,083	22	3,087
UK	253	889	447	1,336
France	138	386	389	775

Source: BTM Consult (2006: 5)

For the purposes of comparison, Table 2.3 presents data on recent capacity growth in the five countries on which this study concentrates. It is worth noting that the Danish onshore market has largely collapsed, whilst in Germany a slow-down in capacity extension is linked to saturation with the best sites taken. Spain too has shown signs of a slow-down. On the other hand, markets are expanding in the UK and France.

Overview of the wind power industry

The wind power industry embraces a range of activities including turbine and component manufacture, project development, operation and maintenance of installations, and associated energy services. The

Table 2.4   **Global market shares of wind turbine manufacturers in 2005**

Company	Accumulated MW 2004	Installed MW 2005	Share %	Accumulated MW 2005	Accumulated share %
Vestas (D)	17,580	3,186	27.9	20,766	35.0
GE Wind (USA)	5,346	2,025	17.7	7,370	12.4
Enercon (G)	7,045	1,505	13.2	8,550	14.4
Gamesa (S)	6,438	1,474	12.9	7,912	13.4
Suzlon (I)	785	700	6.1	1,485	2.5
Siemens (D)	3,874	629	5.5	4,502	7.6
REpower (G)	1,169	353	3.1	1,522	2.6
Nordex (G)	2,406	298	2.6	2,704	4.6
Ecotèchnia (S)	744	239	2.1	983	1.7
Mitsubishi (J)	1,019	233	2.0	1,252	2.1
Others	4,359	567	5.0	4,926	8.3
Total	50,766	11,207	98.0	61,973	105

Source: BTM Consult (2006: 17).[1]

wind turbine industry has achieved exceptional expansion, with an average annual growth rate during 2000–5 of 20.5 per cent (BTM Consult, 2006: 3). Growth in 2005 was a record 40 per cent, taking global capacity to 60 GW, with generation estimated to be 120 TWh (BTM Consult, 2006: v).

Table 2.4 indicates that European leadership in turbine manufacture is clear, with seven firms from the three pioneering countries – Denmark, Germany and Spain – meeting nearly 80 per cent of world demand. However, competition has emerged from other continents. In the USA, GE Energy took over Enron Wind in 2002 and its international expansion is boosted by a buoyant American market. In Asian markets, Suzlon of India and Goldwind of China are set to be major players of the future.

Denmark

The contemporary wind turbine industry has its roots in Denmark. Pioneers in the late nineteenth century – notably Poul la Cour – experimented with electricity generation from wind, whilst in the twentieth century Johannes Juul is renowned for designing the Gedser turbine.[2] During the 1970s and 1980s, wind turbine development was undertaken by green hobbyists, technical experts and small commercial interests, often coming from or selling into the agricultural sector. In Denmark, technological progress resulted from incremental improvement of small machines by interactive learning from the 'bottom-up', and contrasted with the 'top-down', science-push strategy for rapid development of large turbines favoured in the Netherlands, Germany and the USA (Karnøe, 1990; Jørgensen and Karnøe, 1995; Gipe, 1995; Kamp, Smits and Andriesse, 2004).

The result was the 'Danish concept' of wind turbine – the three-bladed, upwind design which has not only assumed market dominance but become an icon of 'green power'. Danish manufacturers came to dominate the sector worldwide by seizing the opportunity offered by the California wind boom of the early 1980s. Although Danish turbines demonstrated superior performance over competitor products, withdrawal of policy support in the USA and the collapse of new investment led to the lean years of 1987–91. The industry went through several waves of company failures, mergers and takeovers. In the 1990s, the five major companies were Vestas, NEG Micon, Bonus, Nordex and Wind World, together accounting for 55 per cent of global wind power capacity installed between 1980 and 1997 (Menanteau, 2000: 243). All of the Danish market was serviced by domestic firms.

Restructuring since the turn of the century has been intense. Nordex transferred most of its manufacturing abroad and listed as a German company in 2001. Vestas and NEG Micon merged in 2003, with the new CEO stating that the aim was to change from a company selling single turbines to one that delivers large power plants, especially offshore (DWIA, 2004). Bonus was taken over by Siemens in 2004. Whilst Danes continue to lead in turbine manufacture and in sectors such as blades with LM Glasfiber, they are dependent on foreign suppliers of components, particularly for gearboxes and bearings. Although market share has fallen, Danish firms remain the major force in the wind industry with over a third of world turbine sales in 2005. Employment creation has been significant, with 20,000 jobs (DWIA, 2005: 6). But as the home market has all but disappeared (to be discussed in Chapter 4) and imported components form a major part of the value chain, the Danish wind industry is almost totally reliant on international markets.

Germany

German firms had a world market share in turbine manufacture of some 20 per cent in 2005, with the major suppliers being Enercon, REpower and Nordex. In the domestic market, Enercon is the dominant company with 43.8 per cent market share in 2005 (BTM Consult, 2006: 69). According to the Association of the German Machinery Industry, half of the global wind industry is located in Germany with total value added of 3800 million euros, whilst 40,000 plus jobs have been created in manufacturing, and another 10,000 in services (VDMA, 2005). These factors result in high aggregate socio-economic benefits arising from wind power. Recognising the industrial and employment gains, government policy to renewables has been highly supportive, as will be seen in later chapters.

Spain

Domestic suppliers dominate the Spanish market. Four firms – Gamesa Eólica, Ecotèchnia, Acciona and M. Torres – installed 72.5 per cent of capacity in 2004 (IDAE, 2005b: 163). Gamesa is the clear market leader, having installed 6000 MW of capacity in Spain – some 60 per cent of total capacity (IDAE, 2006: 5). In the 1990s, Gamesa was part-owned by Vestas and used its technology, but Vestas sold its holding in 2001. The Spanish wind industry is characterised by a high level of vertical integration. Acciona is present throughout the supply chain, since it develops, constructs, operates and maintains wind farms, and also manufactures turbines under the Ingetur brand. Gamesa has two main operating divisions, one for wind turbine manufacture, the other for project development. Other turbine manufacturers also have a project development arm. Dominance by Spanish suppliers of their home market arises from well-adapted products and supportive national policy measures. Wind farms in Spain are often built on hard-to-access hill-crests. Gamesa developed turbines for these conditions, tending to produce smaller turbines than in Northern Europe (Avia Avanda and Cruz Cruz, 2000: 40; Technology Review, 2006: 6).

Spain has a rapidly expanding and energy-hungry economy, which is highly dependent on energy imports. Thus government seized on wind power as an indigenous energy source provided by domestic suppliers as an industrial policy opportunity. Employment creation has been high, with direct employment of 31,750 in design and construction, and 1950 in operation and maintenance (IDAE, 2006: 9). In a country marked by high unemployment, the social benefits of job creation in wind power have been highly valued at regional level and been a driver of the sector's expansion.

France

Since the 1970s, the French solution to energy sourcing challenges and to electricity generation in particular has been nuclear power. France has a large hydropower sector, but other renewables remain underdeveloped. The French wind turbine industry is small and highly specialised. Vergnet produces small to medium-sized turbines, with specialisms in overseas applications; Jeumont (owned by Areva, the French nuclear construction firm) supplies larger machines but has experienced technology problems (Hopkins, 1999; Chabot, 2005). Together Vergnet and Jeumont had a combined share of the French domestic market of 11.8 per cent in 2005 (Chabot and Buquet, 2006: 3–4). The main turbine supplier to the French market in 2005 was the German firm REpower, with a 29 per cent market share. In the same year, Areva took a 21 per cent stake in REpower. This may trigger further investments in manufacturing capacity in France. A number of project development firms exist. One of the largest, SIIF-Energies, has been incorporated into EdF (the major generator from nuclear power in Europe) to form EdF-Energies Nouvelles. The entry of France's nuclear giants into the wind industry brokers a market complementarity between wind and nuclear that is anathema to green movement activists. In time, it may transform the supply chain in the domestic market. In the short-to-medium term, reliance on imported know-how, technology and man-power is set to continue.

United Kingdom

The UK has seen many pioneering initiatives in the renewables sector but few manufacturing firms have survived, mostly due to the lack of supportive policy. Opportunities to promote renewables were missed in the 1970s and 1980s when North Sea oil and gas extraction was the priority, whilst in the early 1990s the policy stress was on propping up an ailing nuclear sector. In addition, Britain's manufacturing base has suffered long-term decline and neglect – whereas in Germany the engineering sector is held in esteem and strongly supported by public policy. These unfavourable conditions have resulted in the near-absence of a national turbine industry, despite the best wind regime in Europe.

Currently, the wind sector in the UK is serviced by foreign firms and imported products. The main suppliers in 2005 were Siemens (67.9 per cent) and Vestas (19.7 per cent) (BTM Consult, 2006: 79). British ownership in turbine manufacture is limited to DeWind. National government has stated the objective of assisting 'the UK renewables industry to become competitive in home and export markets and in doing so, provide employment' (ODPM, 2004a: 4). The UK has sought to attract inward investment, with a little success. Vestas set up facilities in Campbelltown (Scotland); LG Glasfiber manufactures blades on the Isle of Wight. The 'Wind Supply' project aims to promote the entry of British engineering firms into the supply chain, but is in its early stages.[3] However, the UK has not demanded a local production quotient with establishment of manufacturing facilities, as happens in many countries.[4] Although local content demands are problematic under EU treaties, it remains unclear why in relation to renewables Spain and Portugal are able to make them to a greater extent than the UK.

A greater number of firms exist in the wind power services sector, notably in project development. But high concentration and foreign ownership exist here too. Key players include National Wind Power (owned by RWE); Powergen Renewables (owned by E.ON), Falck Renewables (based in Italy), Renewable Energy Systems (owned by the McAlpine group) and Windprospect (independent). With few UK firms in manufacture, employment creation has been low. The DTI (2004: 28–9) estimated a total of around 4000 jobs in the UK wind sector.

The dominance of imported technology and labour, and the concomitant lack of local content and employment, means that the wider socio-economic benefits of wind power seen in Denmark, Germany and Spain barely exist in the UK. Nor are there strong reasons to think this will change. Britain does not have 'national champions' in the electricity sector to whom expansion can be entrusted, as now seen in

France. The British ESI is dominated by foreign capital, with a strong presence of continental utilities notably E.ON, RWE and EdF.

Summary

The early development of the wind industry in Denmark, Germany and Spain gave first mover advantages in terms of technological lead, industry dominance and export opportunities. Economic success contributed to social acceptance as industrial growth stimulated job creation, often in areas suffering unemployment. The pioneer countries in terms of capacity are also those which have the largest domestic industries. National companies effectively control their domestic base, affording opportunities for large-scale export to latecomer countries. The British government has acknowledged that the lack of a turbine manufacturing base poses a problem, but attempts to rectify it have been limited due to a lack of political will and because of the dominance of the market leaders. France has recognised that late entry to the wind sector in European markets can only be secured by shareholding in existing firms and contributing to their expansion. Thus the configuration of the wind industry has impacted on national political sensitivities and the social acceptability of wind power.

Waves of merger, takeover and shareholdings have brought the wind industry closer to the big players in the conventional energy sector. Despite its atypical origins, the growth and transformation of the wind industry is resulting in replication of the characteristics of the wider electrical engineering sector in being large-scale, high technology, capital intensive, globalised yet oligopolistic. Further consolidation and greater dominance by traditional energy firms is likely, as the scale of investment required for the next generation of large-scale wind farms represents a change in order of magnitude which only the energy majors can manage.

Patterns of ownership

Ownership is an important component of the demand side of the equation. National models of ownership emerged in the 1980s and 1990s, but signs of convergence in ownership models have appeared in the 2000s.

Denmark

The early development of the Danish wind turbine industry was stimulated by a distinctive pattern of demand. Installations were typically in the form of solitary turbines or arranged in small clusters. Because of legal requirements regarding residency, owners lived locally. In the 1980s, turbines were mostly installed by wind cooperatives (IEA, 2002: 93), with more than 3000 set up by 1992 (Hvelplund, 2001a: 77). The number of members per cooperative varied between 20 and several hundred people. During the mid-1990s, demand came primarily from farmers installing individual wind turbines (Danish Energy Agency, 1999: 9). This led to a large ownership base, with estimates of total numbers varying between 120,000 and 250,000 individuals (Gipe, 1995: 59; Hvelplund, 2005: 237). With around 80 per cent of capacity owned by individuals and cooperatives, the 'Danish model' was characterised by the predominance of 'private' ownership. The early Danish turbine industry was configured to meet this pattern of demand.

On the other hand, the attitude of the Danish utilities to wind energy in the 1970s and 1980s oscillated between indifference and hostility. As noted by Farstad and Ward (1984: 106): 'wind energy was not seen as a real alternative (...) it was simply considered as basically irrelevant to the interests and activities of the companies'. By the 1980s the threat of new competition, the problems of load management caused by wind power and associated extra costs had been identified by the Danish utilities. Their wish to retain control of the electricity system led to conflicts with the emergent wind sector. Flemming Tranæs, a leading light of the Danish alternative energy movement, recalled that growth in wind power required 'tough struggles, particularly with the traditional monopolies represented by the Danish electrical power companies' (Tranæs, 1996). The government instructed the utilities to invest in wind power, with a first agreement in 1985 and a second in 1990, each for a tranche of 100 MW of capacity (Danish Energy Agency, 1999: 9). The aims were to support technological innovation and to help stabilise the market, at a time when the California wind rush had subsided. Construction of a further 200 MW of onshore wind was imposed on the utilities in 1996 (Hvidtfelt Nielsen, 2005: 113).

Yet by 2000, only 15 per cent of capacity was owned by utilities, whilst individuals owned 59 per cent, cooperatives 24 per cent, and others 2 per cent (Danish Energy Authority, 2002: 6). This pattern of ownership demonstrated significant investment and commitment by a cross-section of Danish society. But what were the reasons for this? The 'Danish model' grew from a social movement that was critical of conventional energy sources and rejected nuclear power. In the 1970s, a top-down initiative by the Danish government and utilities to promote nuclear power was resisted by a bottom-up social movement. The anti-nuclear movement was grouped around the *Organisationen til Oplysning om Atomkraft* (OOA) and from it sprang the *Organisationen for Vedvarende Energi* (OVE), representing the renewable energy sector; these organisations sought alternatives to nuclear, with wind power as the main contender (Farstad and Ward, 1984: 94–5; Jørgensen and Karnøe, 1995). Because of broad opposition, the government initially put the nuclear issue on hold. By the mid-1980s the politicians' appetite for the nuclear option fizzled out. Denmark was unique in that popular opposition prevented the launch of a nuclear program. This had significant consequences. The notion of technology choice by the population – rather than by technocrats – became credible. Opportunities for alternative energy sourcing arose, notably for wind power. Crucially, a lasting association was created between opposition to nuclear power and support for wind power.

What Hvelplund (2002: 66) called a 'green innovation' process was brought about by a coalition of grassroots organisations promoting alternative energy sourcing, including emergent new technology companies, university researchers, sympathetic parliamentarians and a cross-section of the Danish public. Encouraged by spirited activists, the development of the cooperative movement in wind power came to represent a revival of the Danish community spirit (Tranæs, 1996). The ideological reference point for enthusiasts was 'a self-sufficient local community with the idyllic village' (Jørgensen and Karnøe, 1995: 64). But idealism went hand in hand with practicality. As noted by Nielsen (2002: 126) 'people supported alternative sources of energy as part of their determination to create a cleaner environment'. Purchasing, building and repairing wind turbines were means to translate green idealism into practice. Self-help and collective organisation were key components of the movement. The Danish Windmill Owners' Association, formed in 1978, grew out of informal meetings between renewables enthusiasts and environmentalists. Despite resistance from the utilities to these bottom-up processes, the green coalition was able to persuade policy-makers to remove barriers to entry for alternative technologies by supportive policies, including public subsidies. Thus in Denmark local 'ownership' – in both the literal and the extended senses of the term – developed a form of social contract and encouraged acceptance of an emerging and contested technology (Sørensen and Hansen, 2001: 29). But to what extent has the 'Danish model' proved transferable to other countries?

Germany

Across the Danish border in Schleswig-Holstein and neighbouring *Länder* a comparable pattern of small-scale capitalism and local owner-ship of wind power arose in Germany during the 1980s. Renewables activists in the two countries shared similar outlooks, motivations and social movement backgrounds. Indeed, the antinuclear movement and wider 'green movement' in Germany have enjoyed greater political influence than in any other country.[5] Although the protest movement in Germany did not stop nuclear power in its tracks as in Denmark, in 2001 the 'red-green' coalition government decided on a nuclear phase-out. In Germany, too, support for wind power has gone in tandem with opposition to nuclear.

Availability of subsidies for investment in wind farms and guaran-teed feed-in tariffs (discussed in Chapter 4) encouraged a broad owner-ship base, with large numbers of community ventures called *Bürgerwindparks* (citizens' wind farms). In Schleswig-Holstein, which has mainly small and medium-sized wind farms, some 60–70 per cent are citizen-owned. Larger wind farms have been financed through investment funds whose shares have been bought by companies and individuals. Estimates of total numbers of investors vary. EWEA and Greenpeace (2002: 15) stated that over 100,000 Germans owned shares in wind farms. Rickerson (2002) estimated that 90 per cent of turbines were owned by private citizens, with some 200,000 subscribing to cooperative ownership. This pattern of development spread in the neighbouring regions of Saxony-Anhalt and Brandenburg. In strug-gling rural economies – especially in former East Germany – the eco-nomic benefits of wind were especially welcome to small farmers. Although German farmers often own turbines outright, they also lease their land to wind farm operators. These rents are an important source of income (EWEA and Greenpeace, 2002: 15).

To become the world's wind sector leader with over 18,000 MW of capacity, Germany has seen a rapid and intense diffusion process. Increasingly, owners are distant shareholders with commercial interests taking a larger stake. This has been a two stage process. In the 1990s, public fund schemes provided a vehicle for individual investment by German citizens, but loss of tax breaks in 2004 made them less attrac-tive. In the mid-2000s, international institutional investors took a larger share of wind portfolios (May, 2006: 46). On the other hand, German utilities were hostile to wind power, as shown by the PreussenElektra case heard by the European Court of Justice (see Box 2.1) and have not been major players in wind farm construction

Box 2.1　Wind power, the German utilities and the European Court of Justice

In the 1990s, German utilities such as PreussenElektra (now E.ON) and RWE opposed the implementation of the 1991 feed-in law subsidising wind power in Germany (discussed in Chapter 4). Their hostility was explained by the financial burden of paying state-imposed minimum prices for generation from renewable sources, particularly for firms having high levels of wind generation in their grid. Further, existing generators were losing market share to new entrants. Subsequent to proceedings in the German Constitutional Court, PreussenElektra sued Schleswag (a company of which it owned 65 per cent of shares) for repayment of 'additional costs' arising from the feed-in law in the Regional Court of Kiel. The German court of first instance referred the case to the European Court of Justice (ECJ) for a ruling on whether the feed-in law was compatible with European legislation on state aid and on prohibition of quantitative import restrictions. On 13 March 2001, the ECJ ruled that the provisions of the German feed-in law conformed to EC regulations. Reasoning that state aid occurred only through 'advantages granted directly or indirectly through state resources' (ECJ, 2001, paragraph 58), the court rejected the claimed incompatibility with article 87 (1) of the amended EC Treaty since payments incurred came from private, and not public, sources. However in relation to article 28 on quantitative restrictions on imports, the court decided that although the German law involved a trade restriction it was justified on environmental grounds by virtue of the EC's own legislation and with regard to international treaties to reduce GHG emissions, making reference to the 1992 United Nations Framework Convention on Climate Change and the 1997 Kyoto Protocol.

The legal challenge by the German utilities is a clear instance of the 'hegemonic battle' (Elliott, 2003: 185) between conventional and alternative electricity providers. But having lost the 2001 case, the position of the German utilities has evolved. Both abroad and at home, they have sought not to eliminate policy support to wind power but to shape it according to their preferences. In the 2000s, E.ON and RWE built up wind portfolios outside of Germany, particularly in the UK. Further, the ECJ judgement ended legal insecurity regarding the compatibility of feed-in laws with EC treaties and so opened the way for new feed-in laws in Germany and elsewhere in Europe.

in Germany. In summary, despite elements of similarity with the 'Danish model', patterns of wind farm ownership are more hetero-geneous in Germany. The arrival of large-scale wind farms in Germany entailed the need for greater capital and pulled in major financial consortia. In contrast, Denmark, having no large wind farms onshore, has not seen this development.

Spain

With over 10,000 MW of installed capacity, Spain has the second largest wind power sector in the world. However, the dominant pattern of ownership contrasts starkly with Germany and Denmark since util-ities and large firms are the major force. Spanish utilities overcame initial aversion to wind power and became supportive of it. Rather then seeing it as a problem, they came to see it as an economic oppor-tunity. In the diffusion process, a change of industry culture occurred as players copied and learnt from each other. As indicated in Table 2.5, Endesa and Iberdrola, respectively the largest and second largest elec-tricity utilities in Spain, are major wind farm owners. The process has resulted in the emergence of very large players who continue to engage in takeovers and mergers, with Acciona taking over CESA in 2005.

Despite liberalisation, traditional patterns of electricity market dom-inance have continued. As noted by García-Cebrián (2002: 79) 'the utility owning the majority of wind parks in any area is also the utility that has traditionally distributed electricity in the same area'. In the 1990s, regional consortia brought together regional government, the local utility and turbine manufacturers (Avia Avanda and Cruz Cruz, 2000: 38). In the pioneering region of Navarre, the regional govern-ment took a proactive role in setting up Energía Hidroelectrica de Navarra (EHN), with a shareholding of 38 per cent whilst Iberdrola held 37 per cent (Garrigues, 2002: 149). Although EHN had its origins

Table 2.5   **Market share of Spanish wind power developers in 2005**

Iberdrola	32.5
Acciona	10.5
Endesa	8.7
CESA	7.4
Other large firms	23.0
Small firms and individuals	17.9
Total	100

Source: AEE (2006: 47).

in hydropower it become the main player in wind within the region. However, in 2004 it was bought out by Acciona.

In 2005, the four largest developers – all Spanish – owned 59 per cent of all capacity. Small owners in wind power do exist in Spain, but have a marginal place for two reasons. Firstly, competition between the big players for the best sites drives out small investors (Chabot, 2006). Secondly, distributors are reluctant to connect small installations (Blázquez, Calero de Hoces and Lehtinen, 2003: 467). In consequence, wind farms are large, with an average capacity of 27 MW in 2004 (IDAE, 2005b: 161). Thus developers can buy turbines from suppliers in large batches, contributing to market stability for Spanish manufacturers.

France

France is a late-comer to wind power. In the major policy revision of 2000–1 (discussed in Chapter 4), a ceiling of 12 MW per installation was set for eligibility to a favourable feed-in tariff. At face value, the ceiling suggests encouragement of small-scale development on the Danish model. Yet Yves Cochet (2000: 100–1), a key architect of the policy revision, observed that it was difficult to see what logic lay behind the threshold. In practice, wind farms have tended to be large since the 12 MW ceiling could be countered by chopping larger developments into smaller ones. They have rarely involved small investors, because the French regulatory system makes cooperative financing of wind installation difficult. Wind farm ownership in France has been a specialist operation, undertaken by domestic and international project developers.

UK

Wind farm ownership in the UK bears resemblances to the Spanish case in that large firms – often owned by utilities – are the main players. But in contrast to domestic holdings in Spain, UK wind farm owners are generally subsidiaries of foreign firms. National Wind Power Limited – the biggest developer and operator of wind farms in the UK – is owned by RWE, the German utility. Strachan, Lal and Malmborg (2006: 5–6) calculated that over 70 per cent of capacity was operated by 11 firms, many of which were foreign-owned. Community ownership is limited to isolated cases such as the Baywind cooperative in Cumbria. A few farmers own small wind farms, of which perhaps the most famous is in Delabole, Cornwall.[6] Thus the corporate sector, owned by foreign capital, dominates the British scene, creating a pattern of ownership which is the exact opposite of the 'Danish model'.

Summary

In Denmark and Germany, the development of an alternative energy movement led to an ownership pattern which afforded a large place for individual and cooperative involvement, but with limited investment from utilities. In contrast, the ownership base in Spain is dominated by domestic utilities and large industrial firms. But France and the UK did not have as influential an alternative energy sector, nor a well-developed turbine manufacturing industry, nor an engaged electricity supply industry. Hence the preconditions for demand stimulation either by a gradual increase in individual ownership (as in northern Europe) or by major domestic firms (as in Spain) were absent in the 1990s.

In consequence, three ownership models have emerged. The 'Danish model' is characterised by small-scale capitalism and local ownership, with utilities in the background. The 'Spanish model' is characterised by large-scale capitalism and national ownership, with utilities in the foreground. Germany has tended towards the 'Danish model' with a high level of cooperative ownership, though this is not always locally based. France and the UK have *not* followed Denmark but witnessed a third model which is characterised by large-scale capitalism and international ownership, with utilities as owners of subsidiary wind power firms. This variant will be termed the 'international utility model'.

These outcomes put into question the scope for diffusion of the 'Danish model'. The latter has often been advocated as a community model resulting in a high level of social acceptance. The ideological values of the alternative energy movement – environmentally benign sourcing, local participation, embedded generation, decentralised consumption – continue to be rallying points of NGO activists and link with ideals of sustainable development, They have led in the literature to advocacy of a 'community' model of wind power for the UK.[7] However, the 'Danish model' of wind power is not spreading to other countries – because of differences in national political culture and traditions of social organisation, because of changes in technology and industrial structure, and because of current pressures for rapid, large-scale deployment. The factors pushing the sector towards a bulk-power model based on large-scale capital will be developed next.

Changes in wind power deployment patterns

This section explores a paradigm shift in wind power deployment from small-scale to large-scale that has occurred though the interaction between (a) changes in the physical scale of turbines and arrays, and (b) increases in the scale of investment and the consequences for ownership patterns.

Physical scale

The wind industry has seen a rapid upscaling along two dimensions: bigger turbines arranged in larger arrays. The main drivers for upscaling have been efficiency and profitability. Bigger turbines generate more electricity. Larger arrays reduce the marginal costs of construction, grid connection, operation and maintenance. Further, efficiency increases are non-linear, so levering up profitability in major increments. Clarification of these points requires some technical commentary. Firstly, the most crucial element is wind speed (see Box 2.2). Higher wind speed results in higher generation and more income. Secondly, bigger machines are more efficient in terms of 'harvesting' the wind. As noted by the IEA (2002: 60) 'there has been a consistent relationship between machine size and balance of plant costs, with larger machines reducing the cost of the remaining infrastructure on a per-unit-installed capacity basis'.

Turbine size has increased substantially. In 1982, the largest installed turbine was rated at 50 kW. At the turn of the century, the average size of turbines was 1 MW, which represented a doubling of capacity ever five years (Ecofys, 2002: 31). By 2006, the average size of new turbines in Germany was 2 MW (DENA, 2006: 9). The increase in capacity has been achieved by increases in physical dimensions. In 1991–2, typical commercial turbines such as those installed at Delabole or Carland Cross (Cornwall, UK) were rated at 400 kW. Representative measurements were: 32 metres hub height and 17 metres blade making a total height of 49 metres. In the early 2000s, a typical machine was more than three times as powerful at 1.5 MW, and twice the size at 100 metres (67 m hub height and 33 m blade). Where terrain and planning permission allowed, developers in the mid-2000s were installing even bigger machines onshore – up to 3 MW – with machines of 6 MW on trial offshore. It is not known whether the latter will also be used onshore in the future, given the visual and landscape impacts and the difficulties of hauling truck-size components along public roads and up rough tracks.

In the 1990s, turbines also became cheaper. Krohn (1998: 43) noted that the catalogue price for a 600 kW turbine in 1998 was the same as for a 500 kW machine in 1995. In the early 2000s, the cost of installing 1 MW of wind power in continental Europe was typically a million

---

Box 2.2   The importance of wind speed

Wind speed drives not just turbine blades but the whole of the wind power sector. This is because 'the power in the wind is proportional to the cube of the wind speed' (Taylor, 1983: 12). A doubling of wind speed will result in an approximately eight-fold increase in power output. Even a small variation in wind speed converts to a substantial difference in power output. The same turbine on a site with an average wind speed of 8 metres per second (m/s) will produce *twice as much* electricity as on a site with 6 m/s (ODPM, 2004b: 164). The usual cut in speed is 5 m/s (18 km/h) and full-load attained above 12 m/s (50 km/h), whilst the usual cut out speed is 25 m/s (90 km/h). Thus developers expend considerable effort to identify and secure the sites which are most consistently in the optimum range. The differences for yield are substantial: 'operating a plant for one hour at a wind speed of 13 to 25 m/s produces the same electrical energy as operating it for eight hours at a wind speed of 6.5 m/s' (Alt, 2005: 173).

Wind speed is partly a function of height, so placing turbines on hills and on large towers gives access to higher wind speeds. Moreover, power is generated in proportion to rotor area. The larger the area swept by the rotor, the greater the power. A taller tower makes it possible to reach faster winds *and* accommodate a bigger rotor. These factors have driven manufacturers to make ever bigger turbines. The turbines of the mid-1990s swept ten times the area of earlier machines (Gipe, 1995: 154). Over the 1990s, the increase in average capacity of Danish wind turbines from 179 kW to 700 kW resulted in average power production per square meter of rotor area rising from 719 kWh/m^2 to 1,037 kWh/m^2 (Ecofys, 2002: 140).

Wind speed variation has system-wide effects for the electricity generation sector. Wind speed can decrease or increase by a factor of two very rapidly. Each time this happens, generation from a 'wind carpet' – namely, the total number of turbines in a relevant geographical area – decreases or increases by a factor of eight. Fluctuation in wind availability leads to sudden drop-outs and surges in electricity supply, requiring 'up regulation' and 'down regulation' by conventional generating plants. The bigger the 'wind carpet', the more pronounced these effects are. This creates the problem of 'intermittence', which has two different components. The first is the total absence of wind energy – and therefore of generation – during high pressure events. The second is rapid up or down variation in wind speeds and power output. The first can be predicted by weather forecasters with increasing accuracy. Predictions of the second are improving – due to better methodologies and tools (especially under short time-frames) – but will always remain a problem because wind speed variation is inherently a stochastic phenomenon.

euros. However, in 2005–6 turbine prices moved steeply upwards, due to price inflation of raw materials, and a shortage of turbines brought on by rapid expansion in the USA and Europe.

Upscaling has occurred not just in terms of individual turbines, but also their installation. Although the most frequent term to describe an array of turbines is 'wind farm', it unhelpfully masks significant scale differences. Accordingly, a classification in terms of four categories of array is proposed:

1: small installations: stand-alone turbines, or 'clumps' of two or three.
2: medium-sized installations: 'clusters' of four to nine turbines.
3: large installations: 'wind farms' of ten to 49 turbines.
4: very large installations: 'wind power stations' of over 50 turbines.

These category sizes are intended to reflect terminology currently used in the literature, whilst offering a systematic definition of thresholds. A further caveat relates to what is measured. Counting turbines is only one method. Another method is in terms of capacity. A 'cluster' of five 2 MW turbines will have the same nominal capacity as a 'wind farm' of ten 500 kW turbines, though with a smaller 'footprint' and higher generation.[8] But one measurement does not dispense with the other. Site characteristics will place a physical limit on the number, type and capacity of turbines. In consequence, interested parties seek to optimise the combination of turbine number and capacity. Further, average turbine capacities also impact on the equation. Currently the average turbine size is over 1 MW. Consequently an installation today of 50 turbines plus will typically exceed 50 MW nominal capacity. Thus the definitional thresholds proposed reflect meaningful differences between installations.

A trend to ever larger arrays can be found in the UK, where in the 2000s the record for the largest wind power station was being regularly broken.[9] But increase in size of arrays is not a universal phenomenon. In Denmark most capacity is arranged as stand-alone turbines or in clusters: this is an important but little known feature of the 'Danish model'. Wind farms exist, but are few in number. To date, the largest onshore wind farm is Rejsby Hede, in southern Jutland. Comprising forty 600 kW turbines, it was erected as long ago as 1995. The explanation (to be developed in Chapter 7) is that Denmark is a densely populated country and the Danes have been resistant to large arrays. On the other hand, installations of 100 plus turbines were already built during the 1980s in the open expanses of the USA, notably in Altamont, California. In practice, all of the categories are being built in a number of countries. But the key point to recognise is that categories 3 and 4 represent a difference in order of magnitude whose entailments merit exploration.

The consequences of upscaling for the wind sector

Questions of scale impact heavily on investment and ownership, and on construction and operation. In 2004–5, onshore build costs were typically a million euros per megawatt in continental Europe. With wind farms of circa 20 MW then commonplace, each project required 20 million euros (and more as prices rise). But wind power stations of over 100 MW are planned not just in the USA but also around Europe. This scale of investment can only be undertaken by utilities, large companies and financial consortia, reinforcing the 'international utility model' of investment, ownership and operation.

The pervasiveness of this model is illustrated by Table 2.6, which lists the main operators of wind farms around the world in 2005. The listing includes 'traditional' electricity utilities, such as Iberdrola, Endesa, EdF, Vattenfall, RWE and Scottish Power. It includes energy majors such as Shell and DONG, along with new industrial entrants such as Acciona. These companies own wind farms both in their domestic market and internationally. Already each of the two biggest wind farm operators – FPL and Iberdrola – operate wind power capacity greater than in all of Denmark. Further, the global market share of investment and ownership under the 'international utility model' is increasing. As noted by BTM Consult (2006: viii) 'utilities and large energy companies is the dominating customer group in today's markets. They *build* and in many cases, *own* and *operate* the largest new wind farms in the US, Spain and the UK. This customer group owns and operates around 30 per cent of the cumulative installation in the world' (italics in original).

Table 2.6   The leading operators of wind power in 2005

Name of operator	MW capacity owned
FPL (USA)	3,500
Iberdrola (Spain)	3,400
Acciona (inc. CESA) (Spain)	1,700
Babcock Brown (Australia)	1,180
Endesa (Spain)	800
Scottish Power / PPM (UK)	750
EUROS Energy (Japan)	750
Energy E2 / DONG / Elsam (Denmark)	750
ENEL (Italy)	750
Shell Renewables (NL)	740
EDP (Portugal)	740
EdF/EnXco (France)	700
RWE (Germany)	600
Vattenfall (Sweden)	550
Essen (Germany)	480
Total	17,390

Source: BTM Consult (2006: 24)

Further, the large installations favoured by these corporations are increasingly connected to high voltage networks. Thus wind power is losing the characteristics of 'decentralised' or 'embedded' generation which alternative energy enthusiasts valued, and is increasingly used to deliver bulk power. This is the converse of the 'soft path' ideology that characterised wind pioneers and enthusiasts of the 1970s. This development was foreseen by Jørgensen and Karnøe (1995: 73) who predicted: 'the use of still bigger turbines will (...) reinforce the power companies' role in the development of wind turbines shunting aside the alternative energy movement even more'.

In brief, investment and ownership models are causally linked to the scale of wind power deployment. These links are summarised in Table 2.7. Whereas small- and medium-sized installations lent themselves to community ownership and embodied the ideals of the alternative energy movement, putting wind power development on a 'soft' path, today's large installations require investments and expertise that only a combination of utilities and international consortia can provide, so placing the wind sector back on a 'hard' path. With the era of small-scale financing and local ownership largely over, the trend to 'super-size' the wind sector helps explain the controversies that

Table 2.7 Linking scale of wind power deployment with investment and ownership models

Scale	Small	Medium	Large/Very large
Category	Standalone or 'clumps'	Clusters	Wind farms Wind power stations
Investment motivation	Personal needs; hobbyists; green ideals.	Personal investment; sustainability enthusiasts; green ideals.	Business investment; energy diversification; emission caps; profits.
Investors	Individual	Co-operative/ community	Utilities and other large companies/ consortia
Finance and ownership	Local	Local and national	National and international
Industry model	'Alternative energy'; decentralisation; 'soft path'	'Alternative energy'; embedded generation; 'soft path'	Bulk power; 'hard path'.

currently surround it. Upscaling, the accumulation of large wind power stations and the entailments of ownership by international corporations have negative impacts for building social acceptance and will be discussed in Chapter 7.

The challenges of offshore

The development of offshore wind power magnifies these trends and further reinforces the 'international utility model'. By the mid-2000s, the offshore sector was replicating two long-standing characteristics of the electricity industry: (a) the preference of the utilities for bulk power scenarios and (b) the desire to upscale quickly. In consequence, offshore strategy is largely based on wind power stations of 50 to 100 turbines. Some projects are Promethean in their ambitions, such as the proposal for a 1000 MW 'Array' in the Thames Estuary. Yet to date wind farms at sea are few in number, experimental, generally small and near-shore. First generation Danish examples are the 5 MW installations at Vindeby in 1991 and Tunø Knob in 1995. Denmark also has second generation experience, notably Middelgrunden (40 MW), Horns Rev (160 MW) and Nysted (158 MW). The Danish offshore aggregate in 2005 was 398 MW, again giving Denmark pioneering leadership, followed by the UK with 214 MW (BTM Consult, 2006: 39). Meanwhile France and Spain have no offshore experience, whilst Germany has not progressed beyond test platforms. Yet in all five countries large-scale offshore build is considered to represent the future of wind power.

The main advantage of offshore is higher, more constant and more predictable wind speed. Consequently, full-load hours of turbine operation are perhaps 30 to 50 per cent higher. In addition, onshore saturation is already a reality in Denmark and northern Germany, and foreseeable in other countries. Marine settings provide vast expanses for the large-scale deployment that is impossible on land in Europe, whether in terms of size of arrays or turbine dimensions (with 6 MW units on trial and proposals for 8 or 10 MW units on the drawing board). Visual impact is considerably reduced at 8–10 km from shore, whilst turbines at 40–50 km cannot be seen by the naked eye. Finally, the marine renewables sector – wind, wave and tidal – is considered to be interdependent and synergistic, presenting major opportunities for wealth and job creation, especially for nations experienced in offshore energy industries such as the UK and Denmark. These factors led to great optimism in the early 2000s regarding rapid offshore development – with proposals for several gigawatts just for the UK.

However, the severity of the challenges was underestimated. Sea environments are complex and hostile. Conflicts of interests arise in relation to economic uses, such as shipping and commercial fishing, and to environmental desiderata, such as nature and bird conservation. A wide range of competencies are required in engineering terms during construction, but also in scientific terms to understand wind and wave interaction during operation. Offshore installations are considerably more expensive to construct and maintain than onshore, and the sector has a limited track record with some significant failures. Costs vary considerably with location, with distance to shore and water depth being key variables. Interconnection between turbines and connection to onshore grids are more problematic than on land, and costs are high. Because economies of scale are crucial to extract maximum value from long-distance cabling, offshore projects must be large. Apportionment of grid connection costs between developer and grid operator has been a source of contention. Denmark has an onshore 'tradition' of the grid operator paying connection costs, which has been applied to offshore. But in other countries the solutions are hotly debated. Cost estimates for construction found in the literature must therefore be interpreted with caution, due to lack of clarity over the reporting of variables related to location, grid connection apportionment and cost over-runs. Offshore costs have probably been between 30 to 80 per cent higher than onshore.[10] This is a very wide range, and extrapolation of future costs from current levels is problematic.

The risks inherent in marine operation led to a collapse in optimism during 2005–6. For large-scale investment to occur, two core conditions had to be met. One was that the return on investment be attractive: to overcome a funding gap, interested parties sought cast-iron guarantees from governments in the form of capital investment grants and subsidies for generation. The other was to develop suitable contracting structures. The onshore model is typically based on a developer taking the full burden of risk as turnkey supplier. But after the problems experienced in large offshore projects – especially at Horns Rev where Vestas dismantled all its turbines and returned them to shore for repair – no single party is willing to underwrite offshore projects. Multi-contract structures, involving utilities, turbine suppliers and banks, are slowly taking shape for the next generation of projects. But as of 2006, few large projects were moving ahead, though they included Horns Rev II and Nysted II in Denmark, and Barrow in England. The French authorities approved a single project in 2006, whilst German has approved development at Borkum-West. Spanish proposals are still in the discussion stage. Mired in contractual wrangling, price uncertainties, supply chain bottlenecks and inexperience, the offshore boom has been rescheduled to the next decade.

Conclusions

Although the development of offshore remains slow, its rise confirms the major trends in the wind power sector in terms of upscaling, internationalisation and involvement of large-scale capital. On the supply side, turbine manufacture has been consolidated by takeovers and mergers, leaving a small number of large global players. On the demand side, investment and ownership profiles have diverged, with the community-based 'Danish model' contrasting with the utility-based 'Spanish model'. However, the 'Danish model' has been unable to spread beyond northern Europe and is marginalised by the pace of upscaling. The rise of the 'international utility model' probably heralds convergence in investment and ownership patterns in Europe and beyond. The consolidation process seen on the supply side has also occurred on the demand side, with large operators owning increasing proportions of wind capacity. In consequence, the sector is marked by a trend to vertical integration which can cross the supply/demand divide, as the case of Acciona already illustrates. The current and future development of the wind sector is therefore best described in terms of bulk power and as a 'hard path' variant, where generation takes place mostly on the geographical periphery but is centralised in terms of industry structures and ownership.

This development trajectory fits with the practice of ecological modernisation, rather than with the ideals of sustainable development. Barry (2005: 310) identified a 'penchant within ecological modernisation for market-based and entrepreneurial solutions, which turn collective ecological problems for society as a whole into selective economic opportunities for market actors (aided by the state)'. Today's wind power sector provides a representative example of those traits. It forms part of what is termed the 'ecomodernist growth coalition' whose aim is to maintain traditional styles of economic expansion by using new categories of natural resource. This is a far cry from the 'soft path' orientation of wind power's early period, and from the representation of the sector still put about by its partisans. According to the latter, wind power is a community resource, offering a 'hands on', local energy service. In reality, the wind power sector has taken a very different evolutionary turn. To understand why these outcomes have occurred and what their entailments are, later chapters will review and explain the sector's development in more detail.

3
Mobilising for Wind Power

Introduction

Having surveyed the wind sector, we now turn to explanations for the development paths it has taken. This chapter reviews the identities and roles of the actors who have mobilised in favour of wind power, presents their arguments to promote its expansion and identifies their sources of influence. It traces how from heterogeneous starting points different categories of actors have agreed on the common goal of maximising the deployment of wind power and on strategies to attain that goal. The chapter's first section identifies the actors comprising the wind lobby and considers their interests, aims and resources. The second section analyses their cognitive and communicative strategies, whilst the third explores their coalition strategies to mobilise support and resources, and assesses their influence on policy-making.

The major categories of pro-wind actors

The actors mobilising in favour of wind power emerged from a range of backgrounds. We saw in Chapter 2 that the origins of the technology lie mainly with renewables enthusiasts, hobbyists, engineers and small manufacturing firms, who were spurred on in the 1970s by an alternative energy movement resistant to nuclear power and seeking alternatives. But since the 1980s, the amorphous contours of a social movement have been replaced by a more defined and organised set of economic and social actors, characterised by stable relationships with greater institutional underpinnings.

Industry actors and associations

The development of viable markets for wind turbines and the identification of common interests in an emergent and fragile sector led industrialists to form associations at national and international levels. Established in 1981, the Danish Wind Turbine Manufacturers Association brought together major manufacturers (Bonus, NEG Micon, Nordex, Vestas, WindWorld) plus some component firms (Danish Energy Agency, 1999: 19). It is now known as the Danish Wind Industry Association (DWIA). Wind associations emerged in other European countries, notably the AEE in Spain, BWE in Germany, BWEA in the UK and FEE in France.[1] National associations have grouped to form the European Wind Energy Association (EWEA) and the Global Wind Energy Council (GWEC). These are trade and professional bodies who express the voices of the wind industry and promote its interests. Representatives of the wind sector are also prominent in renewables associations such as the SER in France and APPA in Spain. Other categories of federation have also joined forces with wind energy associations. Important examples are the Danish Wind Turbine Owners' Association, which was formed to represent the interests of owners but over time has found common ground with suppliers, and the VDMA – the German Engineering Federation – which is an influential organisation representing all categories of engineering firms in German but which has been influential in promoting the wind power industry.[2]

Separately and sometimes jointly these associations undertake research and analysis, disseminate information, facilitate networking and generally serve the interests of the wind industry. A core strategy is to lobby national and European policy-makers for favourable financial support mechanisms and planning regulations. They also campaign for investment in public research and education related to wind power and other renewables. In short, the wind energy associations handle public relations on behalf of the wind industry and promote the commercial development of their members.

International NGOs

A second group of actors influential in supporting the wind sector are international NGOs, notably Greenpeace, Friends of the Earth (FoE) and Worldwide Fund for Nature (WWF). Whilst WWF focuses on 'hands-on' nature conservation projects, often in partnership with business sponsors, Greenpeace and FoE are environmentalist campaigning associations who organise mass demonstrations and protests,

communicating a political message which is often critical of government and business interests. Because the three organisations are united
by the common aim of combating the disastrous impacts of climate
change, they lobby for policies to cut GHG emissions by reduced
energy sourcing from fossil fuels and increased sourcing from renewables. For this reason, they are consistently favourable to wind power,
joining forces with wind energy associations to lobby government for
supportive policies. As independent campaigning organisations they
work to enhance the legitimacy of renewable energy conversion
technologies as a credible and feasible alternative. Greenpeace and
FoE have been active and visible in supporting individual wind
farms planning applications in the UK. Hence these NGOs play an
important public mobilisation role. To do so, they sometimes collaborate with the wind energy associations, for example with the
'Yes2Wind' campaign and the *Wind Force 12* manifesto, to which we
return below.

Greenpeace and FoE grew out of the antinuclear movement and
have maintained an uncompromisingly hostile stance to military and
civil uses of nuclear energy. Support for electricity generation from
renewables forms part of their programme to phase out nuclear power.
This programme is summarised by the slogan 'goodbye nuclear, hello
wind' (FoE, 2003). On this scenario, wind power substitutes for nuclear
power. Further, in the 2000s, as part of their opposition to so-called
'carbon dinosaurs', FoE and Greenpeace campaigned against coal-fired
electricity generation. But these stances beg difficult questions. Can
wind power effectively substitute for nuclear? If so, with what consequences for GHG emissions? Moreover, can wind power – or even
renewables as a sector – substitute for both nuclear-sourced and
coal-fired electricity? Chapter 6 will return to these questions.

Political parties

The antinuclear, pro-wind stance of the large environmental NGOs is
also shared by 'Green' parties around Europe. Indeed, the membership
base of one category of organisation overlaps with that of the other.
'Green' parties have contributed to putting the nuclear phase-out back
on the agenda, with their most notable success being in Germany. In
June 2001, the 'red-green' federal government entered into an agreement with electricity utilities to place a limit of approximately 32 years
on the operational life of Germany's 19 nuclear plants.[3] It is no coinci-
dence that Germany is also the country with the largest wind power
capacity in the world. In Germany as in Denmark – the two main pio-

neers of wind power – support for wind power has often been accompanied by rejection of nuclear energy.

But this configuration of energy partisanship is far from universal. Although a consensus exists within most European political parties on the need for increased recourse to renewables and improved energy efficiency, political support for wind is not necessarily characterised by rejection of either nuclear or conventional sources of energy. Indeed, in relation to the broad policy goal of promoting renewables, differences exist in the intensity of support, as well as varying preferences in the choice of policy instruments and the targeting of particular categories (wind, marine, biomass, etc). These themes will be developed in chapters 4 and 5.

Institutional actors

A number of institutional actors have been charged with taking forward the renewables agenda at a practical level. The institutional network stems from national governments to embrace energy agencies and public sector research institutions. Examples of energy agencies include DENA in Germany, ADEME in France and IDAE in Spain. The UK has a range of organisations, including the Carbon Trust and the Sustainable Development Commission, together with regional bodies such as the Scottish Renewables Forum. All of these organisations serve to promote renewables, but their precise remit varies. Tasks frequently undertaken are to act as a source of information and as conduit to industrial actors and expertise, to undertake and distribute surveys of developments in the renewables sector, and to distribute subsidies to R&D and to trial projects. The energy agencies sometimes undertake research themselves, but mostly liaise with public sector research institutions and private firms. At the core of the institutional network is the International Energy Agency (IEA) which has an Executive Committee for the 'implementing agreement for co-operation in the research, development and deployment of wind energy systems', a collaborative venture which brings together 18 IEA countries and the European Commission (IEA, 2005: 3). Dating back to 1977, the agreement aims to 'encourage and support the technological and global deployment of wind energy technology', with a particular stress on R&D and information sharing (IEA, 2005a: 15). In discharging these roles, energy agencies have become public champions of the wind power cause.

In summary, the promotion of wind power has been undertaken by actors from different horizons. This broad base of support has been instrumental in converting an emergent and fragile sector into a stable industry. Over time, the organisations identified have found common cause and effectively function as a wind lobby. But the heterogeneity of the origins of the supporting actors also extends to their motivations. The values, interests and aims of the different parties no doubt overlap, but also diverge. This has the potential for disagreements, and needs to be managed by discourse strategies and organisational tactics. To explore these issues, we next turn to the arguments put forward by activists to mobilise support for wind power.

Reviewing the 'story lines' of the wind lobby

The wind lobby has put forward a number of reasons to substantiate why wind power technology should be improved and its deployment accelerated. Their arguments serve not merely to communicate analyses but are used in an advocacy fashion to generate social and economic support. Consequently, the term 'story lines' will be applied to them. The term is not used to belittle, but is employed in a specialised sense. In the formulation proposed by Hajer (2005: 304), 'story lines' are 'the medium through which actors try to impose their view of reality on others, suggest certain social positions and practices, and criticize alternative social arrangements'. Thus 'story lines' fulfil a number of social functions. They provide cognitive structures for comprehension and communication. They allow actors to find affinities and to congregate. A group of actors can refine and amend their 'story lines', converging on issues of mutual concern. In moving closer together, they begin to organise collectively for a common cause. In later and more developed stages, this process can lead to the formation of active coalitions.

In a technological sector such as wind power, 'story lines' also serve to translate complex, technical issues into simple messages that can be communicated to a variety of audiences. In practice, 'story lines' related to renewables help convert scientific and economic expert-oriented discourses into 'down-to-earth' arguments, often with a normative and ethical bent, which are accessible to the public, mobilise support and recruit opinion leaders to a cause. A further move is to convert complex arguments into symbols. A prime example is the three bladed turbine which has become an icon not just for wind power but for the concept of 'clean energy'. By extension, the icon can symbolise a 'brighter future'. Such symbolical usages serve to increase investments of emotional capital, summoning the affective affinities that technical jargon cannot encourage.

For wind power, a number of 'story lines' have accrued. Two core sets will be explored here: one related to environmental issues, the other related to economic issues. These cross-relate to the main policy frames identified in Chapter 1, and as such provide a point of entry for wind activists to lobby policy-makers.

The environmental frame

The environmental frame has offered a reservoir of arguments for promoters of renewable energy sources. Their advocates stress that conventional sources – both fossil fuel and nuclear – have the following negative and unsustainable characteristics:

- they produce pollution and waste;
- are subject to depletion, leading to higher costs in the interim;
- cause insecurity of supply, because of the need to import from producer countries prone to political tensions.

On the other hand, renewables are presented as having the following positive and sustainable characteristics:

- they are 'green', 'clean', and 'friendly', since in use they produce no atmospheric pollution or hazardous waste;
- are inexhaustible and 'free', leading to stabilisation of costs;
- increase security of supply, due to indigenous production.

These characteristics add up to what is sometimes called 'ecological sustainability'. This is not the same as 'sustainable development' as outlined in Chapter 1, since it relates to only one of its three 'pillars'.

These markers of sustainability contribute towards a 'story line' which can be subsumed with a term often heard in RES discussions, namely 'intelligent energy'.[4] This 'story line' communicates a particular view of reality in suggesting that it is energy sourcing – rather than energy use – which is smart, bright and clean. It downplays technological progress in the handling of conventional energy sources to achieve greater efficiency and better environmental performance than was possible in the past. It also adds a normative dimension towards technology choices in implicitly characterising conventional energy as 'dumb' and castigating its use. The 'story line' thereby allocates positions to social actors, criticises present practices, and offers alternatives. It also opens a pathway to influence policy. As noted by the IEA (2002: 36) 'much of the current market for wind energy is principally driven by the very low life-time emissions that the technology offers'. Policy-makers are increasingly concerned with emission reductions, and the 'intelligent energy' 'story line' promises a fast route to attain them.

This approach is particularly marked in relation to the 'fight against climate change'. Wind energy associations and environmental NGOs have teamed up to position wind power as a core technology to help meet GHG reduction targets agreed under the 1997 Kyoto Protocol. In the *Wind Force 12* manifesto for wind power, the EWEA and Greenpeace (2002: 12) claimed that 'a reduction in the levels of carbon dioxide being emitted into the world's atmosphere is the most important environmental benefit from wind power generation'. In a list of the 'global benefits of wind power', the GWEC and Greenpeace (2005: 10) made the bold claim that wind power 'reduces climate change'. More specifically, these organisations claim that:

> currently wind power installed in Europe today is already saving over 50 million tonnes of CO_2 every year. In terms of carbon delivery, wind energy is outperforming many other proposed solutions. The European Wind Energy Association business-as-usual target for 2010 of 75 GW, a doubling of installed capacity in 6 years, would deliver one third of the EU's Kyoto commitment. (...) wind power is one of the few energy supply technologies that have the maturity, clout and global muscle to deliver deep cuts in CO_2 (GWEC and Greenpeace, 2005: 5).

At this juncture, I will not dwell on the substantive content of these assertions.[5] Rather attention is drawn to how a climate change 'story line' is built up. Based on the valid observation that wind turbines do not emit greenhouse gases, a claim is made that their deployment leads to emission reductions from sources that do. This accords with the *values* of environmentalist NGOs concerned by the disastrous effects of climate change. It also accords with the material *interests* of industrialists, who gain a powerful sales argument to market their products. This distribution of potential benefits allows coalition behaviour between the often antagonistic categories of environmentalists and industrialists. Further, the 'story line' allows an injection of urgency into deployment rates for wind power by setting imminent target dates for substantial expansion. In the translation of more gigawatts of wind capacity into fewer tonnes of CO_2 we find an instance of how technical information can be used 'in an advocacy fashion, that is, to buttress and support a predetermined position' (Sabatier and Jenkins-Smith,

1993: 218). The imposition of a particular 'view of reality' – as Hajer (2005: 304) called it – is abetted by the assertion that few other means are available to meet the Kyoto deadline. In brief, the wind lobby 'story line' on climate change serves to create a common cause between disparate actors, generate policy relevance, inject urgency and disqualify alternatives.

The economic frame

The economic advantages and challenges related to wind power have given rise to several 'story lines', notably on the themes of 'security of supply', 'affordability' and 'David versus Goliath'.

Security of supply

According to Andrew Garrard 'the key selling point for wind power has moved from the environmental argument to security of supply'.[6] Renewables offer several varieties of security of supply advantage. The problem of depletion does not arise for renewably energy, which is inexhaustible. Renewables are indigenous and so are not subject to geopolitical risks such as severance. In reducing reliance on imports, they improve self-sufficiency, help with the balance of payments and accord with ideologies of 'national independence'. Whilst total displacement of conventional sources by RES is currently impossible, they permit energy diversification. Further, the 'fuel' is 'free' in the sense that energy from the sun, wind or water is not traded on markets. This contrasts with fossil fuels, which are prone to price volatility. In consequence, hydro and wind sources display greater price stability and long-term predictability. These sources also require no fuel transportation infrastructure or costs. In the post '9–11' climate of fear, a twist has been added to the 'security' theme by suggestions that decentralised installations, such as wind farms, are less sensitive to terrorist attack than large, centralised facilities (especially nuclear power stations) or exposed oil and gas pipelines. This combination of geopolitical and economic arguments stresses the contribution renewables can make to a diversified and balanced energy portfolio. A telling example of the use of the 'security of supply' 'story line' was the EWEA 'no fuel' campaign of 2006.[7]

Affordability

'Security of supply' discourse connects with the 'affordability' argument that wind power is a competitive, cost-effective and necessary component of sustainable energy policy. The medium to long-term aim of the wind lobby, as set out in *Wind Force 12* and similar documents, is to implement a 'high-renewables/high wind' scenario, in which wind power plays the role of a 'path finding' technology for RES-E in general. Its claimed route to attain this end is to substitute wind power for coal-fired and nuclear power stations nearing the end of their life. Wind lobbyists are at pains to argue that the life-cycle costs of wind are lower than those of nuclear and other rival sources (Milborrow, 2004). The GWEC and Greenpeace (2005: 2) stated that wind power 'delivers the energy security benefits of avoided fuel costs, no long-term fuel price risk, and wind power avoids the economic and supply risks that can (sic) with reliance on imported fuels and political dependence on other countries'. The argument stressed that the 'fuel' cost for wind is zero, unlike fossil fuel installations. In the context of rising prices for oil and gas in 2005–6, the zero fuel cost of wind is presented as a hedge against inflation. A complementary argument is that the costs of integration of wind power into electricity supply systems – for grid extension and back-up power of various kinds – are relatively low (Milborrow and Harrison, 2004). Advocates stress the benefits of rapid deployment, since wind farms are modular installations which can be built more quickly than conventional power stations, especially nuclear. The benefits of job creation, regional development and export potential are frequently cited.

However stress on the 'affordability' of wind power can strike a dissonant chord with the argument that wind power remains at the 'near-market' stage and still requires subsidies. Thus Richard Stark of ILEX Energy Consulting acknowledged that 'as an industry we need to decide whether we're campaigning that our costs are coming down and we can compete with nuclear and gas, or whether we need to be campaigning that we need renewable support mechanisms for a long time' (quoted by Massy, 2005: 53). In more precise terms, this comes back to asking what *levels* of support are appropriate, for how long and under what conditions – themes to which we return in the next chapter.

'David versus Goliath'

'Story lines' serve not just to exploit opportunities but also to manage threats. One of the greatest arises from competitors. Elliott (2003: 185) referred to 'the conviction that many environmentalists share that the powerful fossil fuel industries will inevitably seek to marginalise rivals like renewables and, unless challenged, will continue to enjoy the strong backing of governments, to the detriment of renewables and conservation' and termed the ensuing contest as a 'hegemonic battle'. For many renewables activists, the battle is waged against both the fossil fuel and nuclear sectors. To communicate positions and choices in this contest, a 'David versus Goliath' 'story line' is often aired in wind power contexts.

The ambition of the wind industry is to set itself up as a major source of electricity generation. In mature European markets characterised by low increases in demand, its hunger for growth can only be satisfied by taking market share from rivals. Competing sources that are targeted are coal and nuclear. Thus in response to the 2002–3 UK energy review, the BWEA (2002) argued that 'wind can meet the nuclear shortfall', and urged the government to increase policy support to allow wind to replace the 7 per cent of national generation from nuclear sources scheduled to disappear by 2010. More typically, however, wind power replaces generation from fossil fuels and thereby can achieve cuts in emissions. But the question of whether wind power should target the displacement of nuclear or coal (assuming for the moment that the choice exists) divides the wind lobby. The environmentalist NGOs who mobilise in favour of wind power wish to phase out nuclear, offering wind as an alternative. On the other hand, wind industry representatives are neither unanimous nor categoric in rejecting nuclear. Electricity majors such as British Energy, Areva and EdF deal with both nuclear and wind power. Likewise component manufacturers sell into a range of markets, whilst the careers of electrical engineers typically embrace different conversion technologies, including nuclear. In the industrial context, stress falls on complementarity between energy sources, rather than on conflict as it does in the environmentalist NGO context.

Viewed from a different angle, the 'David versus Goliath' 'story line' expresses fears shared by environmentalists and at least some industrialists. One fear is that, on a roughly ten year time horizon, the nuclear sector is well placed to capitalise on the need for GHG reductions (since it is virtually carbon free at the point of generation) and that it is poising itself to do so by undermining renewables. In the UK during the 2000s, a variant of this scenario was that the wind industry is 'on trial' and has a limited window of opportunity to prove itself. During the 2005–6 Energy Review, the marked interest displayed by Prime Minister Blair in nuclear power seemed to give support to this interpretation of the government's strategies.

The 'David and Goliath' 'story line' surfaces in other contexts too. The hostility of the German utilities, manifested in the case brought by them before the European Court of Justice (see page 33), is an historically important example of the 'hegemonic battle'. But even in mundane operational terms, the wind industry considers itself disadvantaged since procedures are designed for conventional generating technologies and may cause unfair treatment. In this vein, Milborrow and Harrison (2004: 35) asserted that 'hundreds of amendments to market regulations are being made, but wind, forced in the main part to abide by rules designed for thermal generators, is a particular victim of the process'. The 'fear of Goliath' syndrome may also explain the tendency of wind activists to consider their critics as either disguised representatives of the nuclear industry or climate change deniers serving the interests of the fossil fuel majors.

An advantage of this 'story line' is that in casting David as a tiny and defenceless victim, aggression is justified against Goliath. In other words, the 'David and Goliath' characterisation provides a strategy that is both defensive and aggressive. It is defensive in that it portrays rivals – notably the nuclear industry – as powerful and unscrupulous preda-tors. It is aggressive in that the expansion of wind power is expected to be at the expense of competitors. Thus the 'story line' orchestrates responses to economic and competitive challenges. It provides a means of justifying and advancing the cause of wind power – sometimes bel-ligerently – but it also communicates worries about the future, moti-vating activism on the basis of fear. However, a problem with this 'story line' is that its relevance varies with national contexts. In Denmark and northern Germany, investors are indeed small and often community-based (as explored in Chapter 2). Historically, they have often felt threatened by the big utilities and display a 'David' mental-ity. However, in the UK and Spain the same 'Goliath' corporations own both renewable and conventional installations.

In summary, the main functions of these 'story lines' are to promote the interests of the wind sector and communicate a favourable image of it to governments and the public through the provision of:

- a means to conceptualise the role of wind power within the electric-ity generation sector;
- a justification for its rapid deployment on the basis of both environ-mental values and economic aims;
- a foundation on which to build coalitions among interested parties to influence policy and optimise deployment conditions.

From 'story lines' to coalition building

Through the medium of these 'story lines' a number of actors have drawn together to achieve common purposes. The nature of their cooperation is perhaps best described as a 'discourse coalition', defined as 'a group of actors that, *in the context of an identifiable set of practices*, shares the usage of a particular set of story lines over a particular period of time' (Hajer, 2005: 302 – italics in original). A core example of industry-NGO partnership is the regular publication of *Wind force 12* by the EWEA/ GWEC and Greenpeace (2002, 2005). This document sets out an expansionist manifesto for the long-term future of the wind industry, offering periodic reviews of progress towards its goals. In the UK, practical examples of inter-organisational link-ups include the joint communication on wind farm development and nature conservation put together by English Nature, RSPB, WWF-UK and BWEA (2001).

To promote wind power, NGOs have provided legitimacy through discourses on sustainability and climate change which emphasise the 'common good', whilst corporations have provided technical expertise and financial backing. Without the backing of international NGOs, private, profit-orientated firms would find it difficult to stake claims about acting in the public interest. Simultaneously, the credibility of NGOs claiming expertise in energy and climate policy is enhanced by association with industrialists offering a technical solution. The complementary relationship between NGOs and the wind industry has received additional support in those countries (Denmark, Germany) where community ownership – with its stress on values comparable to those of the NGOs (such as local participation and equity through shared costs and benefits) – has prevailed.

Coalitions between firms and NGOs are characterised by an *exchange of resources*, whose dimensions embrace the technical, the financial, the relational and the communicative spheres. Each party benefits from the exchange, but on distinct and separate bases. NGOs benefit in non-material terms by enhancement of their reputation and influence (and perhaps in material terms by increased capability to attract sponsorship), whilst firms benefit in material terms (though they can also draw non-material advantage by a 'greening' of their credentials). Significantly, such coalitions do not have disputes over the distribution of benefits. This is because the rewards of one party do not subtract from the other. Indeed, the multiplication of rewards for one can trigger advantage for the other. This positive sum game makes coalition behaviour between NGOs and firms an attractive proposition. However, a threat emerges from the 'clash of fundamentally different worldviews' (Arts, 2002: 34) which arises when NGOs lock themselves into narrow industry-oriented proposals which restrict their wider remit as societal critics. Further, claims of civil society associations to define the 'common good' can be compromised if they are perceived to be promoting private interests. Indeed, in the wind power debate the tension between the public interest and private gain never lies far beneath the surface and has led to objections by critics (discussed in Chapter 8).

On the basis of common 'story lines' and resource exchange, a loose confederation of pro-wind actors has evolved into a well-orchestrated lobby. In so doing, pro-wind action has moved from *ad hoc* arrangements to being increasingly purposive and goal-directed. The building of solidarity and shared ethos has proceeded along two parallel tracks. On the one hand, technical and technological issues have been converted into economic goals in order to mobilise the interests and resources of players in the electricity industry. On the other, environmental issues have been converted into normative and ethical goals in order to reflect the values and mobilise the resources of NGOs. These twin tracks allow influence over public opinion and access to the political domain.

Mobilising opinion

A key example of joint working between industrialists and NGOs in the UK is the creation of vehicles to mobilise public opinion and encourage a social contract favourable to wind power. As flagged by Toke (2002: 100): 'the British wind power industry needs to find ways of connecting up with the grassroots so that wind power supporters feel they have an incentive to argue for specific schemes. In this way the enthusiastic opponents of wind power schemes can be counterbalanced by enthusiastic supporters of the technology'. The wind lobby has responded by the creation of the 'Embrace the revolution' and the 'Yes2Wind' campaigns, two initiatives which reflect the perceived need to popularise wind power in a social climate sometimes characterised by opposition.

Launched in September 2004, the BWEA 's 'embrace the revolution' campaign has an unashamedly propagandist slogan which reflects the electricity industry's long-standing and self-serving ideology of progress.[8] The campaign has relied on a web site, public exhibitions and celebrity endorsements to convey an up-beat message on the environmental advantages of wind power. The aim has been to demonstrate that national popular opinion favours wind power and so counter local opposition. To this end, great stress is placed on opinion polls. These are commissioned regularly and headline results – of the variety that 80 per cent of respondents are in favour of wind power –

appear prominently on the web site. A key aim has been to encourage 'local champions of wind' near proposed sites to organise petitions and canvass politicians. The campaign presents an example of the 'public relations' approach to wind power in the UK. It is characterised by a 'top-down' communicative strategy which contrasts with the bottom-up involvement of community and cooperative actors in Denmark and Germany. The risk with this approach is that it promotes passive acqui-escence rather than active participation.

The 'Yes2Wind' campaign has been organised by a consortium bring-ing together BWEA, Greenpeace, FoE and WWF. It is a variety of 'hearts and minds' campaign to rally public support for wind power. Under the slogan of 'a clean, energy future', it mobilises at the local level, seeking to galvanise support for specific wind farm planning applica-tions and counter antiwind opposition during the planning process. The campaign's name is a manifesto in itself. The 'yes to wind' propo-sition communicates an unconditional acceptance of wind. This is the mirror image of the unconditional rejection of nuclear, which has been part of the identity of Greenpeace and FoE. The unconditional rejec-tion of one energy source (nuclear) combined with the unconditional acceptance of another (wind) is an unusual configuration of opinion. It has generated high levels of partisanship in recent energy debates. It contrasts with a core principle of the planning regime which is that decisions to accept or reject applications are made on the basis of defined criteria (discussed in Chapter 7).

Securing policy influence

Cooperation between industrialists and NGOs has been directed towards improving the policy environment for wind power. Corporate actors wield major resources, not only due to their economic and financial power but also in terms of knowledge and expertise. In condi-tions of liberalisation, such as those applied to the ESI in the past two decades, central government has become more dependent on the market, not just for delivery of goods and services, but for the informa-tion on which to form policy decisions. Thus Smith (2000: 103) argued that 'the structural position of industry as the source of information and organisation resources bestows privilege and influence'. Obtaining a de facto monopoly of expertise is an important goal. In this, cor-porate actors are helped by two structural factors. One is that firms are the originators of information on technologies and techniques, drawn from a unique experience base. The other is that as pri-vate market actors they enjoy considerable financial and institutional independence to undertake R&D, do market surveys, canvass opinion and influence policy.

New and emergent technologies are a case in point. The main sources of knowledge are technology purveyors. At the initial stages of the learning curve, the operational conditions of the technology – by which is understood not just technical characteristics, but also its insertion into economic, industrial, social and physical contexts – are known only partially. When a technology is rapidly evolving, so too are its conditions of operation. Almost invariably the technology purveyors are private firms whose vested interests determine their strategy towards the transmission of information. Thus they jealously guard their patents and other knowledge monopolies. They are selective in the information they provide to policy-makers and the public, whilst such information as is volunteered suffers from rapid obsolescence. Firms are skilful in evoking reasons why they should not be more transparent, such as 'commercial confidence', the stresses of competition with established providers, the unfairness of existing institutional or operational parameters, or the vulnerability of nascent sectors and new firms. At the same time, requests for information provision can be turned and crafted for the purposes of advocacy in order to modify the perception of policy problems, to predispose solutions towards preferred outcomes, to court public opinion, to change incentive structures, to protect existing market share or ring-fence new markets. Access to policy-makers, through governmental committees and advisory boards, provides opportunities to guide policy in the industry's preferred direction. Conversely, in their dealings with technology-driven industries, policy officials rely on information from industrial and corporate interlocutors. Officers of state can round out their knowledge by other means, including commissioned research from independent consultancies or universities, parliamentary commissions and public inquiries. However, the scope to call in such information is limited. Consultation exercises tend to privilege the economic actors who already enjoy a superior reputation or resources.

Moreover, decision-makers become dependent on private sector actors not just for policy information, but also delivery. International agreements give lobbyists opportunities to increase such dependencies on the part of central government decision-makers. In the 1997 Kyoto protocol, signatory governments committed to GHG emission cuts. One readily available pathway to reductions was assumed to lie in the expansion of the renewables sector. The more optimistic the assumptions regarding GHG reductions from the expansion of renewables, the more governments became dependent on the associated technology purveyors. In many countries, the main beneficiary of this dynamic was the wind industry. A similar process occurred in relation to EU directive 2001/77/EC (discussed in Chapter 4) which aimed to increase the proportions of electricity generated from renewables in member states. Both of these agreements set near-term targets, for the 2008–12 commitment period in the Kyoto protocol and 2010 for the EU directive. By locking themselves into imminent deadlines, governments became inextricably dependent on private oligopolies for the delivery of public policy targets.

Conclusions

A heterogeneous coalition of actors, including industrialists, NGOs and politicians, has assembled to promote wind power on the basis of 'story lines' regarding energy supply and climate change. These 'story lines' have dove-tailed with the premises of 'ecological modernisation' and its stress on 'win-win' scenarios to attain high environmental and high economic performance by technological innovation. They provide the cognitive and communicative conditions to rally around the slogans and icons of 'clean energy'. International energy and climate policy reforms provided a propitious content for pro-wind action groups to move from *ad hoc* arrangements to being increasingly purposive and goal-directed. The pooling of the political resources of NGOs (and affiliated politicians) with the economic resources of the corporate sector has allowed the wind lobby to access policy-making circles. International agreements gave the wind lobby the opportunity to induce dependencies on the part of central government decision-makers, who relied on technology purveyors to deliver against politically determined targets for RES-E generation and GHG emission cuts. Over time, the wind lobby's policy advocacy became increasingly focused on 'large-scale penetration' of wind power – in other words, maximisation of market share – in a context where key competitors (coal and nuclear) were destabilised. The major goal was to enhance policy support mechanisms, such that they allow profitable investments in wind power. Associated goals have been to lobby for more permissive planning regimes and improved grid access. Detailed analysis of support mechanisms will be undertaken in Chapters 4 and 5, and associated themes will be developed thereafter.

4
Promoting Wind Power through National Policies

Introduction

The development of wind power in the recent period has displayed signs of both success and fragility. Its expansion has not been the result of a market-led dynamic nor of a societally-driven process (except to an extent in Denmark and Germany). On the contrary, its deployment has been the result of political process, since its development has required supportive public policy initiatives. This chapter will analyse the manners in which this has occurred. Its aim is to analyse policy choices at the national level, identifying the major innovations and assessing their outcomes. It reports critically on the extent to which policy-makers have reviewed and improved national policy on the basis of experience over the long term. Core questions addressed are: How far has policy been interventionist or 'market-centred'? What are the development path consequences of particular choices of policy design? The first section provides a policy overview by reviewing the justifications for support to renewables and setting out the main options available. In the second section, the historical development of policy instruments to promote wind power is analysed in relation to Denmark, France, Germany, Spain and the UK. Chapter 5 will then concentrate on cross-national comparison in other to evaluate the rich variety of experiences and draw policy lessons.

Policy overview

Justifications for policy support

The need for policy support to renewables has arisen from a number of factors.

Firstly, the 1990s and early 2000s were a period of plentiful and cheap conventional energy sources, resulting in low electricity prices. This made it hard for alternative energy sources to compete. However, this era is drawing to a close. The 2005–6 period saw sharp rises in oil and gas prices, with a return to glut and rock-bottom prices unlikely.

Secondly, conventional sources have often been subsidised. Historically, the coal extraction industry has received considerable public subsidies in European countries. The nuclear industry received and continues to receive major subsidies for R&D.

Thirdly, the 'external costs' for the environment and society associated with conventional energy sources are not fully included in market prices. The 'polluter pays' principle, enshrined in article 174 of the Treaty on European Union, has been imperfectly applied to electricity supply. The combination of subsidies and unpaid 'external costs' kept prices of conventional fuels artificially low. However, 'externalities' are increasingly being factored into the equation. The major new instrument to achieve this is the European Union Emissions Trading Scheme (EU-ETS), which is a market-based mechanism designed to set a price for carbon.

Fourthly, in the post-liberalisation phase, electricity generation from conventional sources was on the basis of 'sweating assets' in a context of over-capacity. In other words, the aim was to extract lowest prices from equipment which was often ageing and fully depreciated. However, low prices from 'sweating assets' cannot be sustained indefinitely. Investment in *any* form of new capacity to fill the 'generation gap' and meet future needs will raise costs.

Fifthly, electricity industry structures have presented obstacles to the deployment of renewables. The development of new technologies in a context of vested energy interests and market dominance by international companies seeking to protect market share was inherently problematic.

Sixthly, at market prices renewable energy sources have generally been more expensive than conventional sources. To overcome this handicap and encourage technological innovation, sources have been subsidised. During its early development, wind power consistently needed subsidy to compensate for economic handicaps. But by the 2000s this situation was evolving in complex manners. Although wind power became increasingly competitive with conventional sources, its competitiveness varied greatly due to systematic variations in wind speed. According to Ecofys (2002: 30–1) 'at the best sites, the costs of wind energy are directly competitive to fossil fuel (...) the price range for onshore wind energy is from 5 (best UK sites) to 9 c€/kWh (German sites)'. Hvelplund (2005: 239) calculated that the cost of producing wind power in Europe varied from 3 c€/kWh on the best coastal sites (such as in Ireland), to 7 c€/kWh on good inland sites on mainland Europe. For the UK, reported generating costs for onshore wind varied in the range 3.1 to 5.4 p/kWh, whilst for conventional sources the range was 1.9 to 5.2 p/kWh (The House of Commons Environmental Audit Committee, 2006: 43).[1] Further, during the mid-2000s conventional fuel prices were rising, due to market pressures and to the effects of carbon pricing within EU-ETS. Meanwhile renewables prices were falling. The ensuing phenomenon of price curve convergence has been particularly significant in the case of wind power, due to the falling costs of wind turbines combined with increased productivity due to upscaling. In consequence, wind is now considered to be 'near to market' on many sites and price competitive on the best. This means that the level of policy subsidy needed to compensate handicaps is closely related to site quality: onshore, the level has reduced in most instances and subsidy is not necessary at all in some scenarios. Under these circumstances, policy-making is highly complex.

Establishing the need for support to renewables is thus only the initial component of policy-making. The core decisions relate to the *nature* and *level* of support. Further, the level of support needed has changed over time and continues to vary by country and location, as the national case-studies will show. The requirement in the early period was to provide substantial subsidies to enable the emergence of a new technology. But in the current period, the need is to design 'cost reflective' subsidy schemes which ease wind power into the market ready phase.

Policy design

Howlett (2002: 246) offered the following classification of policy components: the ends of policy-making can be divided into abstract *policy goals* and concrete *policy specifications*, whilst the means can be divided into choice of *instrument type* and alterations in *instrument setting* (such as setting the level of a tax or subsidy). To convert this classification into an everyday comparison, a choice of instrument type is like a *vehicle* whilst policy goals can be compared to a *destination*. Changes in instrument setting allow the driver – the policy-maker – to *steer* the vehicle to its destination. The *goals* of energy policy are typically the provision of supply at competitive and affordable prices, the assurance of energy security and the reduction of environmental externalities.

In relation to renewables, these goals have translated into a philosophy of 'more is better'. As put by Haas *et al.* (2004: 834) 'the main focus [of wind policy] must of course always be to trigger investment in new capacity'. This is a common starting point, and taken in isolation could suggest a reductionist view of policy. However, it opens out onto associated objectives including the promotion of technological progress, increased competitiveness, industrial policy and social welfare. With renewables, the key policy *specifications* concern the regulation of *prices* and / or *quantities* (e.g. how much new capacity is built). Further specifications may concern locational choices, categories of investors targeted and response to changing market circumstances. Because most of the wind policy debate has concentrated on identifying a fast-track to achieving production capacity increases, the most common specification has been to set national capacity targets.

The EU has taken a keen interest in the deployment of renewables, with the main policy measure on RES-E being directive 2001/77/EC (often referred to as the European Renewables Directive). This established an EU target of 22.1 per cent of electricity consumption to come from renewable energy by 2010. It enjoined member states with obligations to publish reports on progress towards targets, issue guarantees of origin of electricity produced from RES, give guaranteed access to the grid, ensure transparent and non-discriminatory connection costs and reduce administrative obstacles. National targets are indicative and not binding,[2] and are presented in Table 4.1 for the five countries surveyed. They relate to *all* categories of RES-E, with no breakdown specified for individual sources.

Table 4.1 RES-E targets in the European Renewables Directive

	RES-E 1997 TWh	RES-E 1997 %	EU RES-E 2010 targets %
Denmark	3.21	8.7	29.0
France	66.00	15.0	21.0
Germany	24.91	4.5	12.5
Spain	37.15	19.9	29.4
UK	7.04	1.7	10.0
EU	338.41	13.9	22.0

Source: *Official Journal of the European Communities*, 27.10.2001, L283/39

The targets specify market share, but in practice policy specification based on percentage targets proves vague. The electricity supply industry operates on the basis of operational *capacities* (measured in megawatts or gigawatts) providing quantities of *generation* (measured in megawatt hours or terawatt hours). However, it is difficult to translate percentage targets into a quantity of generation because total future electricity demand cannot be predicted with precision. Even when a 'guesstimate' is made for total generation, an aggregate target for *all* RES-E technologies begs many questions. A more meaningful form of target-setting is to specify a quantity of generation from a specific conversion technology. Once a generation target is identified, then a capacity target can be estimated. Yet here too, calculation problems arise. Renewables such as hydro and wind are weather and site dependent, leading to considerable year-on-year variability in output. Over a period of time, the difference between 'good' and 'bad' years can be of the order of 30 per cent. Yet indicative targets for both capacity and generation in relation to each technology are still desirable because without them the scale of investment cannot be estimated. Identification of investment costs on a comparative basis facilitates choice between rival technologies. However national practice in the specification of targets varies considerably. In Denmark, Germany and Spain, policy-makers established clear targets for wind power, but in France and especially the UK targets have been ambiguous. One consequence is that the real meaning of percentage targets – and progress to targets – has to be painstakingly reconstructed by analysis. Yet more troubling is the entailment that policy-makers may lack a precise idea of what their goals are and what the costs of reaching them will be.

The Renewables Directive did not specify policy instruments – so allowing considerable variation in practice around Europe – but did leave scope for harmonisation at a later date. Thus the choice of *instrument type* and especially its *setting* is made by national governments, but is open to lobbying by interest groups. As a large range of instruments have been used to support wind power, a typology of policy instruments will be set out to clarify the national case-studies.

Policy instrument typology

The range of instruments available for the promotion of renewables has received extensive commentary in the literature.[3] Consequently only a brief introduction is offered here, whilst detailed examples are included in the national case-studies below. 'Technology-push' instruments develop technologies, improve products and assist with commercialisation. They include R&D programmes and credits, as well as approval and certification schemes. 'Demand-pull' instruments stimulate markets. They include subsidies, price support, taxation and regulations (for example, related to grid connection and planning). Subsidies can be *direct* (financed by taxation) or *indirect* (paid *via* consumer bills). They have two main destinations: capital investment or production. The main forms of capital investment subsidy are direct grants and tax exemptions. Production subsidies in electricity generation take the form of a payment per kilowatt hour. These measures mostly target commercial operations. Measures directed at consumers include campaigns to raise awareness, incentives for green energy consumption, and net metering.

As regards wind power, the most important instruments have been production subsidies embodied in tender schemes, feed-in tariffs and quota systems. With a tender scheme, the government decides on a quantity of capacity which is allocated to market actors on the basis of their bids. As a generalisation, tender schemes are designed to drive prices down. But for RES-E, allocational criteria are not based solely on price, as environmental and/or social criteria can also receive weighting. With feed-in systems, a tariff per kilowatt hour is imposed by government or agreed by stakeholders, and the market decides on the quantity (e.g. the level of capacity to be developed). These systems are termed renewable energy feed-in tariffs (REFITs). Their proponents stress that because income is guaranteed, risk is reduced and so REFITs draw in a great range and number of investors. With quota schemes, government fixes a quantity of generation (equivalent to a market share) for RES-E and issues 'green certificates' in relation to units of generation. Certificates are tradable and constitute a revenue stream for RES-E generators. In principle, the price of 'green certificates' is set by supply and demand, but in practice their value is determined primarily by interventionist mechanisms associated with quota-setting, and only secondarily by the market. These schemes are often termed 'renewable portfolio standards' (RPS). Supporters of RPS schemes argue that they encourage competition between technologies and are more 'market compatible' than REFITs. All of these schemes provide revenue for RES-E generators. But revenue is not enough. A final and crucial component of the policy regime relates to previsions for taking power into grid. This is crucial for wind power, given the need for grid connections at remote locations and for balancing services. Previsions concern the availability of grid connection and accompanying contractual arrangements, notably the questions of who pays and for what. As will be seen in the case-studies, all these parameters vary considerably.

National case-studies

Since the 1980s, policy choices related to wind power have been refined on the basis of 'learning by doing'. To explore the outcomes, four main themes will be addressed in relation to each of the national studies:

1) evolution of policy content;
2) reasons for policy choices;
3) distinctive features of each national regime; and
4) the flexibility of support systems and the scope for improvement.

Denmark

In the Danish political system, consensus is important and is reached on the basis of cross-party 'political agreements'. As stressed by the Danish Ministry of Environment and Energy (1996: 9): 'Denmark has a long tradition of implementing vigorous energy policies with broad political support and the keen commitment of a wide range of actors: energy companies, industry, grassroots, municipalities, research circles and consumers'. These factors facilitated the use of indicative planning in the energy sector.

Following the Brundtland report, the Danish Ministry of Energy (1990: 89) linked its *Energy 2000* action plan to achieving sustainable development, placing emphasis on climate change and GHG reductions. In prioritising environmental concerns, it moved away from the security of supply arguments dominant in the 1970s following the oil crisis. The plan set targets for 2005 for 10 per cent of primary energy consumption to come from RES, with 7–8 per cent to be achieved by 1995. The 1996 action plan, entitled *Energy 21*, reaffirmed environmental commitments with a 20 per cent reduction of CO_2 emissions from their 1988 levels targeted for 2005, and a halving by 2030 (Danish Ministry of Environment and Energy, 1996: 3). The main means were energy efficiency and increased recourse to renewables (Danish Ministry of Environment and Energy, 1996: 73). A target was set of 12–14 per cent of energy consumption from RES in 2005 and 35 per cent by 2030. In relation to wind power, targets were set of 1500 MW onshore by 2005 and 4000 MW offshore by 2030 (Danish Ministry of Environment and Energy, 1996: 76). These aspirations were matched by a highly supportive policy regime.

Policy design

Denmark has been a pioneer in renewables policy. Experiments have resulted in major successes – notably, the emergence of the wind power sector and the creation of a world-class turbine industry – but also mis-judgements in choices of policy instrument and their settings. Analysis will be divided into three periods: the expansionary phase of the 1980s and 1990s, the transition phase of 1999–2001, and a consolidation phase post-2001.

Wind power deployment was encouraged by a range of subsidies. Investment credits to wind turbine purchasers started at 30 per cent in 1979, were progressively reduced to 10 per cent and abandoned in 1989 (Hvidtfelt Nielsen, 2005: 110). But credits jump-started the turbine market. Users assessed the relative merits of different makes, and compiled a comparative database in *Naturlig Energi,* the publication of the Danish Wind Turbine Owners' Association (DWTOA). This encouraged interactive learning between turbine producers and buyers, facilitating incremental improvements and upscaling of the technology (Karnøe, 1990; Kamp, Smits and Andriesse, 2004). Investment came mainly from private investors, stimulated by tax breaks (Tranæs, 1996; Hvelplund 2002: 72). The utilities invested in wind power only when pressured by government, with 100 MW in 1985 and another 100 MW in 1990 (Danielsen, 1995: 60). Forced investment by the utilities was an industrial policy measure to support turbine manufacture in a lean period for exports.

Production subsidies proved to be the most supportive instrument. Guaranteed payments for electricity sales to the grid evolved into what are now called feed-in tariffs. In 1984, the DWTOA negotiated an agreement with the utilities fixing the remuneration for wind generated electricity at 85 per cent of consumer prices (Heymann, 1999: 125). When negotiations on its renewal broke down, the government drew up legislation in 1992 on feed-in tariffs and associated issues. Between 1992 and 1999 tariffs were again set at 85 per cent of household prices (minus charges and administrative costs), on top of which the government paid a direct subsidy of 0.17 DKK/kWh and a 0.10 DKK/kWh reimbursement of the Danish carbon tax. According to Redlinger, Dannemand and Morthorst (2002: 205), the total price in the mid-1990s was 0.6 DKK/kWh, approximating to 7–8 c€/kWh.[4]

In this period Danish wind power policy had the following features:

- stable price support, combining payments from the utilities with state subsidy;
- a guarantee of 'prioritised' dispatch to the grid for independent producers;
- a sharing of grid connection costs with utilities that was broadly favourable to producers.

The predictable price regime – underwritten by legislation – and preferential grid access provided a viable basis for contracting loans from financial institutions by individuals and cooperatives. A major expansion in wind power resulted, rising from 9 MW in 1980 to 343 MW in 1990, and to 1442 MW in 1998 (Danish Energy Agency, 1999: 30). Investors secured high profitability, since the relative generosity of the system compensated risks associated with an emergent sector and untried products. But by the late 1990s, the technology was well-tested and the acceleration in new build raised questions over ballooning subsidy levels,[5] as well as over the optimal configuration of the wind sector. Strong pressures for policy reform developed.

The 1996 EC directive on the liberalisation of European electricity markets raised questions over the future of the Danish feed-in system. The 1999 Danish Electricity Supply Act transposed the directive and restructured the ESI by an unbundling of generation and grid ownership. This entailed reform of support to renewables, given complaints from the utilities over the burdens it placed on them in open markets. The act initiated a shift away from subsidised feed-in tariffs to a new quota system based on 'green certificates', scheduled for introduction in 2000. Minimum and maximum certificate prices were set, fixing support in the range of 0.10 to 0.27 DKK/kWh (Danish Energy Agency, 2001: 6–7).

Proponents argued that the quota system was in greater conformity with the market principles driving the liberalisation process and would reinforce competition. At that time, the European Commission was thought to favour EU harmonisation of renewables support schemes using quota systems. In the Preussen Elektra case then before the ECJ (see page 33), the German utilities had attacked the feed-in tariff system. The Danish authorities made the assumption that REFITs would be ruled an illegal state aid. According to Meyer and Koefoed (2003: 604): 'the Danish government wished to go in front in order to influence the operational rules of the model that it believed would be the future choice for the EU'. Denmark was looking for a 'first mover' advantage and to be a winner in 'regulatory competition' at the European level.[6] A second argument related to the cost to the state budget of subsidies at ever-higher wind penetration levels. Other criticisms related to excess subsidy to some turbine owners.

A coalition to oppose the certificates scheme was formed by the DWIA, the DWTOA, the Confederation of Danish Industries and the OVE. Critics complained that the interests of existing turbine owners were threatened, and that new investment would stop due to political uncertainty. Hvelplund (2001a: 108) argued that the quota was unworkable because annual fluctuations of 20–30 per cent in wind availability would produce unacceptable price oscillations, as well as making it sensitive to price manipulation. The DWIA (2002: 8) claimed that 'an isolated Danish market will be so small, so complex and so risky that it most likely will collapse'. A 'boom and bust' situation arose, with record installations of 606 MW to beat the year 2000 deadline, and a market nose-dive to less than 100 MW in 2001 (IEA, 2002: 44). The heated debate led to postponement of the proposal, and a transitional scheme was put in place. Once the ECJ confirmed the legality of the German feed-in tariff in March 2001, the Danish government had a way out. With the collapse of domestic orders for wind turbines, the Danish Minister for the Environment and Energy abandoned the quota system in September 2001, but left the transitional scheme caught 'between state and market'.

The change of government in November 2001 settled the uncertainties. The new right-wing government was highly critical of wind power, and took the opportunity of the 'green certificates' debacle to slash subsidies. The political agreement of 19 June 2002 released consumers from the obligation to purchase wind-generated electricity, with claimed savings of 2 billion DKK up to 2008. Danish wind tariffs became the lowest in Europe, with a maximum revenue of 6.5 c€/kWh in 2005 (Agnolucci, 2006). Their structure is highly complex, with different rates paid according to date of installation and whether privately or utility owned.[7] Its main features are that:

- an environmental premium is paid on top of the wholesale electricity price;
- the combination of 'price plus premium' is capped;
- a ceiling on eligibility exists, defined in full load hours of operation;
- older turbines retain high priority dispatch to the grid, but new ones must have market contracts;
- premiums are paid by consumers (no state subsidy).

Consolidation has been effected at two levels. Existing turbine owners continue to be compensated, but with a gradual phase-out of feed-in tariffs (Danish Energy Authority, 2005: 23). New investment is encouraged albeit to a limited extent, with a mere 4 MW of additional capacity in 2005 (IEA, 2006a: 101). As of 2006, Denmark has a 'repowering' target of 350 MW of onshore capacity.[8] In addition, 400 MW of new offshore capacity was programmed on the basis of calls to tender. The combination will take the 2005 aggregate of 3128 MW close to

4000 MW by 2100, and help meet Denmark's target of 29 per cent of electricity from RES-E. This should not be difficult, as RES-E was already at 28 per cent in 2004 (Danish Energy Authority, 2005: 4). Thus the low growth rate needs to be set in a context of targets met, onshore saturation, planning difficulties, as well as grid and market integration problems. Later chapters will develop these points, but firstly the main outcomes of the Danish experience will be reviewed.

Policy discussion

Denmark gained and lost leadership in wind power, but retained its role as pioneer. The major achievement of the expansionary phase was the development of the feed-in tariff. Danish policy-makers combined an innovative policy instrument with national technological traditions, planning policies and dispersed ownership structures to foster the development of the new industry of wind turbines. The result was a gradual process of technology diffusion, with close controls on access to subsidies and ownership being preferred by supporters of wind power, administrative authorities (including land-use planners) and the utilities, who saw regulation as a means to contain a competitor. Synergies developed between an alternative energy movement and an innovative manufacturing sector, mediated by a consensus-seeking political system. R&D credits fostered the creation of a centre of excellence at the Risø research institute, with major outputs being the European Wind Atlas (Risø National Laboratory, 1989), wind modelling software, and turbine safety testing and certification procedures. To use the DWIA's favourite metaphor, Denmark became the world 'wind power hub'.

The political will to maintain leadership was the catalyst for the quota system proposal. But it was wrong-footed by the slow pace of European liberalisation, by a 'multi-speed' Europe in which member-states and their industrial 'national champions' pursued individual agenda, and where core institutions (the European Commission and ECJ) were out of phase. The 'green certificates' scheme was not only out of time in relation to the European concert, it was out of tune with the Danish constituencies that made the wind sector a success. The botched transition indicated the importance of 'path dependence'. Once policy routines are embedded in the institutional landscape and the livelihood of influential actors comes to depend on their persistence, the scope for major reform is severely constrained. The political lesson is all the more stark as the 1999 reform was *not* the product of neo-liberal ideologues, but emerged from a political agreement broached by a broad left government. The lesson of 'path dependence' was not lost on other countries, encouraging them to retain existing renewables policies whilst diminishing the European Commission's zest for harmonisation on an RPS basis.

In the current consolidation phase, the challenges of improving the integration of large-scale wind power into the Danish power supply are being addressed. With wind power in the mid-2000s contributing 15–18 per cent of electricity supply, the previous *ad hoc* arrangements for grid and market integration reached their limits. Reforms included an obligation for new wind plants to sell their output directly on the power exchange. Grid management issues are now significant and will be discussed in Chapter 6.

In summary, the Danish approach has been innovative, multi-faceted and largely consensual. But in being pioneering, the Danish 'laboratory' has been a source of policy learning, rather than producing 'ready made' models.

Germany

Even though the wind resource is low with few onshore sites having wind speeds above 7.5 m/s (Rickerson, 2002), highly supportive policy initiatives enabled Germany to take over leadership in the wind power sector from Denmark. Although policy has been similar, Germany has systematised the support mechanisms to a greater extent and – being a large economy with a far greater land mass – has installed considerably more capacity.

Early German policy to renewables majored on R&D support (Ibenholt, 2002: 1183). The GROWIAN project of the 1970s produced a 3 MW turbine but failed to achieve commercial feasibility. However, private firms in both Denmark and Germany developed viable turbines in the sub-500 kW class.[9] Support was then reviewed and the stress placed on demand-pull policies. The '100 MW Wind Programme' was established in 1989 and enlarged to 250 MW in 1991, supported by a production subsidy of around 3 c€/kWh (Rickerson, 2002; Bechberger and Reiche, 2004: 49). These measures catalysed the building of the first wind farms, of which some also benefited from financial support by regional governments (Badelin, Ensslin and Hoppe-Kilpper, 2004: 7). By 1990, wind power capacity was 56 MW (BMU, 2005: 12). These initiatives stimulated a manufacturing base and encouraged use of feed-in tariffs.

Policy design

The 1991 law (the *Strom-Einspeisungs-Gesetz*) set the feed-in tariff for wind and solar at 90 per cent of average electricity prices, with a pre-1999 price of circa 8 c€/kWh (Wüstenhagen and Bilharz, 2005). An important innovation was to impose a purchase obligation on the regional utility monopoly. The policy regime created a low risk environment for investors, since it guaranteed returns and gave long-term stability. Meanwhile additional system costs, such as the provision of balancing services, were transferred to the utilities.[10] Soft loans for capital investment were available from public sources (Bechberger and Reiche, 2004: 50–2). The result was a wind boom, with capacity rising from 98 MW in 1991 to 4444 MW in 1999 (BMU, 2005: 12).

Yet problems were encountered concerning the structure and level of tariffs, the absence of differentiation in relation to wind speeds, and geographical disparities in the distribution of wind farms (Grotz, 2002: 116). Historically, regional companies held distribution monopolies, but the concentration of wind farms in northern Germany resulted in disproportionate impacts on grid operators and customers (Ringel, 2006: 6).[11] Further, liberalisation of the electricity sector in 1998 left consumers free (in principle) to change suppliers. The pegging of feed-in tariffs to consumer prices proved problematic. Although fairly high initially, tariffs fell in value due to price reductions subsequent to liberalisation. Hence a reform of the feed-in regime was needed to redistribute costs on a national basis.

The Renewables Energy Sources Act (*Erneuerbare Energien Gesetz – EEG*) of 2000 incorporated several innovations. To favour climate protection, the 2010 target was to double the share of RES in energy consumption. The bill also aimed to reduce energy imports, improve security of supply and encourage technological development (German Parliament, 2000). The law obliged transmission system operators (TSOs) to purchase and sell all RES-E produced, with an equal proportion of RES-E to be incorporated into the electricity mix of all suppliers, thereby sharing the economic burden. Arrangements for grid costs were also clarified, with the developer paying for connections and the operator financing grid upgrading. Complaints from energy intensive industries over RES-E costs led to a capping of their prices, with consumers making up the difference.

The effectiveness of the feed-in tariff in promoting investment was demonstrated in the 1990s. By 2000 the challenge was to encourage efficiency. To improve the competitiveness of renewables, price differentiation was reinforced:

1. between technologies (with photovoltaic receiving the largest pro rata subsidy);

2. between sizes of plant (in relation to hydro, biomass and so forth but *not* wind farms);
3. in terms of yearly tariff reductions; and
4. in relation to average wind speeds.

Indexation of tariffs on average prices was abandoned. For 2000–1, the wind tariff for onshore sites was set at an initial rate of 9.1 c€/kWh, reducing to a minimum rate of 6.19 c€/kWh. All installations were entitled to the initial rate for five years, and to relevant tiered rates for a further 15. The fall to the lower rate was triggered once a 'reference yield' threshold was passed. Remuneration on high wind-speed sites dropped more quickly than at low wind-speed sites. This favoured the dispersal of wind farms. It also enabled evaluation of the performance of turbine models and promoted competition.

An important innovation was the principle of 'degression'. This involved a tariff reduction of 1.5 per cent every year for new installations. In consequence, wind farms built later would receive lower levels of return. 'Degression' aimed both to reflect greater productivity and to increase competitiveness, in a period when turbine prices were falling. In programming a reduction in economic incentives, the measure stimulated build by offering higher tariffs to early starts. This proved a successful measure: capacity increased from 6112 MW in 2000 to 16,629 MW in 2004, with output soaring from 9500 GWh to 25,000 GWh (BMU, 2005: 12). The EEG paid lower prices yet still ignited a second wind boom.

The amendment of the EEG in 2004 set a higher generation target of 12.5 per cent for RES-E by 2010 – transposing European directive 2001/77/EC – and added a 20 per cent target for 2020. Incentives for 'repowering' were introduced whilst tariffs for offshore installations were increased, but only to wind farms outside of nature conservation areas. Payments to new onshore wind farms were reduced to 8.7 c€/kWh for the first five years, dropping to a minimum level of 5.5 c€/kWh. To be eligible, sites had to meet at least 60 per cent of the reference yield in order to 'quash any economic incentive to install wind turbines on sites with poor wind conditions' (BMU, 2004b: 8). The annual 'degression' increased to 2 per cent. These changes were designed to tighten support, squeeze out excess profits and encourage price convergence between renewable and conventional electricity sources.

Policy discussion

The German REFIT proved a flexible and adaptable policy instrument. Recourse to systematic review demonstrated a willingness to recalibrate the instrument, and avoid perpetuation of privilege (or disadvantage). It also showed that policy review and price adjustments are possible without jeopardising investor confidence. A remarkable feature of the German case is the linking of energy policy and industrial policy, with environmental policy providing the connecting ground. The aim was not just to generate more electricity, but to renew the German manufacturing base, enhance engineering expertise and create jobs whilst improving environmental performance. The policy frame was explicitly based on 'ecological modernisation'. Federal Environment Minister Trittin stressed the need for 'an "ecologically optimised" development path for renewable energy sources' (BMU, 2004a: p. viii) through reconciliation of environmental, economic and technological objectives. German leadership in the wind sector emerged from a multi-dimensional and long-term approach to policy-making which not only diversified energy sourcing, but also stimulated technological and industrial development, opening market opportunities at home and worldwide.

Spain

The key drivers in Spanish energy policy have been economic growth and national 'independence' in sourcing, with environmental considerations some way behind. To help attain energy policy aims, Spain has retained a tradition of indicative planning. The 1999 Renewable Energy Promotion Plan restated the key objective of reducing Spain's high level of energy dependence of 72 per cent in 1998 – compared to an EU average of 50 per cent (IDAE, 1999: 1). With RES-E (mainly large hydro) at 20.3 per cent of total generation in 1998, ambitious goals were set for renewables, including a target for wind power of 8974 MW of capacity for 2010 to generate 21,538 GWh (IDAE, 1999: 77). Whilst the capacity target was exceeded early with 10,000 MW by 2005, generation at 20,955 GWh fell short of expectations (CNE, 2006: 10). Nevertheless Spain has the second largest wind power sector in Europe.

In the context of continued, rapid growth in Spanish energy demand, high dependence on imports (80 per cent in 2004) and the GHG emission targets of the Kyoto Protocol, the 2005 Renewable Energy Plan presented renewables as a means to diversify energy supply, reduce the trade deficit and stabilise the economy (IDAE, 2005a: 73). The target for renewables was to attain 12 per cent of primary energy consumption in 2010 (in comparison with 6.3 per cent in 1998), including 30.3 per cent electricity from RES-E (IDAE, 2005a: 5–17). An estimated 95,000 new jobs were expected through successful implementation, with 37,000 in the wind sector (IDAE, 2005a: 78).

These ambitious targets were premised on a highly successful feed-in policy, developed incrementally over the long term as in Denmark and Germany.

Policy design

In the early 1980s, Spanish wind power policy was largely limited to research funds from the Ministry of Industry for experimental turbine designs. Subsequently, support came from national feed-in tariffs and regional investment subsidies. The roots of Spanish renewables policy lie in the 1980 Energy Act, which responded to the second oil price hike by seeking greater energy independence and improved efficiency. This led to a policy framework known as the 'special regime',[12] developed progressively through a series of legislative provisions.

The 'special regime' was institutionalised in a 1994 royal decree, setting out a differentiated subsidy system for the various categories of RES-E. An obligation was placed on regional distributors to purchase RES-E at a guaranteed premium price for a five year period. This developed a feed-in system comparable to the 1991 *Strom-Einspeisungs-Gesetz*. However, no long-term guarantee on prices was given, with yearly revisions made by ministerial order. The decree was amended by the 1997 Electricity Industry Act on the liberalisation of electricity markets, setting wind tariffs in the range of 80–90 per cent of retail prices and guaranteeing grid access. The act triggered a wind power boom. Whereas capacity stood at 227 MW in 1996, it doubled each year during 1997–9, reaching 2288 MW in 2000.

The new pricing mechanism was more than an 'inflation peg'. Over time, it was developed into a flexible coupling with market prices. The 1998 decree gave a choice of payment options to generators: a fixed price per kilowatt hour or the 'market option', based on the average pool price of electricity plus a premium per kilowatt hour. The 'market option' came in two variants: one involving offers to the spot market, and the other not involving offers but simply tracking prices. In this phase, wind generation firms moved increasingly to the latter variant (and away from fixed prices) as total revenues were higher (Bustos, 2005: 21).

Prices and premiums continued to be fixed annually by the Ministry of Energy and Industry, and the decree required that the tariff system be reviewed every four years. Thus prices were controlled by national government, with limited transparency and predictability in their setting. Distributors were required only to sign five year contracts. Despite no guarantee of renewal thereafter, developers seemed not to attach risk to the arrangement (Dinica, 2002: 218). At the informal level the permanence of the scheme was taken as assured, even though at the formal level the early 'special regime' displayed significant uncertainties. These features favoured 'insiders' and large firms, notably the utilities. In consequence, small investors have only a limited presence in the Spanish wind power sector, unlike Denmark and Germany.

The 2004 royal decree on the 'special regime' situated its support of renewables within the context of achieving sustainable development. It promulgated two key developments. Firstly, it increased the predictability of the 'special regime' by making its methodology more transparent. Payments are indexed on the Average Reference Tariff (ART), which in 2005 stood at 7.3304 c€/kWh. Although the ART is set yearly by government, it reflects market prices. The institutionalised linkage with wholesale prices reduced administrative intervention, stabilising the 'special regime'. Secondly, it reinforced the market option. It ruled out the variant of tracking prices without making offers, but gave economic incentives to active spot market participation.

The decree continued to guarantee revenue support *via* two main options – a regulated tariff and a market option – with producers opting into a scheme for a 12 month period (but thereafter can choose again). The regulated tariff option is available to installations under 50 MW, but is modulated in relation to capacity and year of start-up. The generator sells direct to the distributor, and receives 90 per cent of the ART. In 2005, the average price paid was 7.108 c€/kWh (CNE, 2006: 9). In the market option, the generator makes offers to the spot market and receives the pool price plus subsidies minus penalties. The subsidies are 1) a premium, fixed at 40 per cent of the ART; 2) a market incentive, fixed at 10 per cent of the ART; and 3) a small payment for guarantee of power. Wind farms above 10 MW must make production forecasts for the day ahead. A penalty is imposed if deviations from forecast exceed 20 per cent (as compared to 5 per cent allowable in the 'ordinary regime'). The cost of deviation is 10 per cent of the ART. In 2005, the average net payment for wind power was 8.661 c€/kWh (AEE, 2006: 68–9), rising in line with wholesale prices to 10.125 c€/kWh in the first semester of 2006 (CNE, 2006: 9).

The drawbacks of the market option for wind farm operators are that guarantees on income are lower, whilst extra costs are incurred through the requirement for forecast schedules and penalties for deviations. The latter are especially a problem for small installations (AEE, 2006: 35). Nevertheless, most operators have migrated to the market option because it pays better. With revenues some 25 per cent higher than in the regulated tariff, the proportion of operators choosing it rose from 20 per cent to 93 per cent in the course of 2005 (AEE, 2006: 70). By 2006, only the smaller producers – with an average installation size of 5.5 MW – were still in the regulated tariff (CNE, 2006: 9). Further, Spain's wind power boom continued with a record rise in new installations of 2201 MW in 2004 alone, bringing the aggregate to over 10,000 MW by 2006 (CNE, 2006: 10).

Policy discussion

The usage of REFITs in Spain proved distinctive. The design and setting of the policy instrument provided the economic incentive necessary for the major expansion in wind power programmed in indicative planning documents. But Spanish policy-makers broached the transition to a market-oriented regime early and deepened this approach to a greater extent than in Denmark or Germany. The transition was made easier by the fact that the major recipients of feed-in tariffs in Spain have been the utilities, not private investors. As established commercial actors, Spanish utilities offered less resistance to a market transition, which has been softened by economic incentives. Unlike Germany, the Spanish system does not have a 'degression' element. On the contrary, it encourages generators to migrate to a market-based scheme, where prices have tracked fuel price inflation for conventionally sourced electricity. High prices under the market option are attractive for large firms, but the costs to society have been high. In the early 2000s, the Spanish authorities apparently believed this was worth paying in the context of extremely high national dependence on world energy markets. However, escalating gas prices meant that by 2006 the tracking of conventional energy prices led to excessive 'windfall' profits. A reform of the 'market option' was being undertaken in early 2007 to address this problem.

France

Like Spain, France is short of conventional energy resources and has sought to increase national independence in energy sourcing. Whilst this priority guided policy-makers to nuclear power, it generated less impetus in relation to RES-E. Although France generates a greater quantity of electricity from renewables than any other EU country (with the exception of Norway), this is almost exclusively from the large hydro projects undertaken in the post-war period. France has given little support to other renewables. In the 1990s, a tender scheme entitled

Eole 2005 set a target of 250–500 MW of wind power by 2005 (Benard, 1998; Laali and Benard, 1999). It was based on the UK's NFFO (discussed below), and had similar disappointing outcomes. Four rounds of calls to tender between 1996 and 2000 produced low bid prices of around 5 c€/kWh, but from the 360 MW accepted only 125 MW was operational by 2001 (Chabot, 2001; Cochet, 2000: 112–13; Menanteau, 2000). The scheme was abandoned due to lack of interest caused by the stop-go process, administrative complexity, high risk of rejection and limited attention to environmental and social factors (Cochet, 2000: 41).

Policy design

The year 2000 Electricity Act established a new framework for renewables policy. It created a dual system for wind power, with feed-in tariffs for installations below 12 MW and a tender system above. Modelled on the German EEG, the French feed-in system as set out in the ministerial order of June 2001 had the following features:

1. It set regulated prices (the 2001 tariff for continental France was 8.38 c€/kWh for the first five years).
2. It obliged the transmission system operator to buy electricity generated by eligible installations.
3. It incorporated the principle of degression into the fixing of the tariff.

Degression applied in three ways: (1) a yearly reduction of the tariff of 3.3 per cent after 2003; (2) a further reduction of 10 per cent once 1500 MW of capacity was installed, these lower rates being applicable to new build and (3) a 'price tier' system was applied, whereby the same initial tariff was payable in all cases for the first five years, but for the following ten years the tariff would be calculated in relation to output. Rates were calculated using a sliding scale based on full-load hours (using an average of three years of the first five, discarding the best and the worst years). Up to 2000 hours, the rate remained at 8.38 c€/kWh, dropping to 5.95 c€/kWh at 2600 hours, and to 3.05 c€/kWh for 3600 hours (and above). Tiered pricing favoured dispersal to lower wind sites and discouraged the 'wind rush' phenomenon of excessive concentration in high wind-speed areas. The goal of the French system has been to achieve 'fair and efficient' tariffs, such that 'on medium quality sites (in France approximately from 6 m/s at hub height), a minimum profitability is possible, and so that on high quality sites (…) the profitability may be higher but not undue'

(Chabot, 2001: 336). Tight eligibility thresholds were set. Only installations up to a maximum capacity of 12 MW qualified for guaranteed prices and associated advantages. The 1500 MW threshold triggering price revision was a unique feature that raised questions over long-term development.

The French government was slow to put out tenders for wind farms above 12 MW, but in 2004 calls were made for 500 MW onshore and 500 MW offshore (DGEMP-DIDEME, 2004). Companies tendered a total of 457 MW onshore, of which seven projects with a cumulative capacity of 278 MW were accepted; the average tariff being 75 €/MWh (Ministère de l'Economie, des Finances et de l'Industrie, 2005a). Ten tenders for offshore were put forward of which only one project of 105 MW was accepted, at a tariff of 100 €/MWh (Ministère de l'Economie, des Finances et de l'Industrie, 2005a). Although little explanation was given for these decisions, bid prices were clearly important.

Many parts of France, particularly coastal areas, have good wind conditions. This, combined with favourable government policy, can be expected to produce a large increase in wind power. Faster rates of build have occurred since the introduction of feed-in tariffs, but not on the scale of the German or Spanish booms. With 366 MW of new capacity, 2005 was a record year for construction and took the cumulative total to 918 MW in May 2006.[13] Wind power output of one TWh in 2005 was equivalent to 0.22 per cent of national consumption which stood at 482 TWh (Chabot and Buquet, 2006: 5). But with total generation from RES-E at 12.8 per cent of consumption in 2005, France had a lower share of renewables in electricity supply than in 1997, when it stood at 15 per cent. Yet France's 2010 target is 21 per cent. If France is going backwards, it is because hydro output has dropped due to weather conditions and tightened environmental controls, whilst other renewables have not come on-stream. Indeed, early translations of the 21 per cent target into capacity were well off the mark. In 2003, a ministerial decree set targets for January 2007 of 2000–6000 MW of onshore wind power, 500–1500 MW offshore, with other RES-E in a range of 567–1810 MW (République française, 2003). Yet even the interim 2007 targets as stated could not be met. Further, the true scale of expansion required to meet the 2010 target is considerable higher. Early estimates of a need for 10,000 MW of wind power capacity as made by Cochet (2000: 114) and Boston Consulting Group (2004: 15) were more realistic. By the mid-2000s, the slow rate of build signalled the need for policy review.

The 2005 French Energy Bill reinforced support to renewables, whilst programming a renewal of conventional energy sourcing, including nuclear power. It transposed EU directives and targets on RES-E and biofuels, made proposals for solar heating, biomass and energy efficiency certificates, and set a target of 10 per cent of all energy sourcing to come from renewables by 2010 (the 2005 level was approximately 6 per cent). The feed-in system was preserved and simplified. The 12 MW ceiling on eligibility was abolished as of July 2007, abandoning the 'dual system' with its calls to tender. The tariffs were amended by the decree of 10 July 2006 which set the initial rate at 8.2 c€/kWh for onshore wind turbines in mainland France.[14] Although this looks like a reduction relative to the 2001 level of 8.38 c€/kWh, lifetime revenues will increase because the 'degression' elements have been substantially softened. Whereas the higher base rate was paid only for five years under the 2001 decree, in future it will be paid for ten years (falling thereafter for a further five years to between 8.2 and 2.8 c€/kWh, depending on the windiness of the site). In addition, the year-on-year reduction in tariffs will be 2 per cent (previously 3.3 per cent) and is scheduled only beyond 2008, whilst the aggregate 1500 MW ceiling for triggering lower tariffs has been abolished.

Policy discussion

What then are the causes of the slow rate of build? One important question is whether the tariffs are too low. Financial incentives were decreasing in a context of increasing turbine prices in 2005–6. The 2006 change in the tariffs instituted a more generous level of remuneration, whose consequences have yet to work through. However, other problems also exist including problems with grid connection (discussed in Chapter 6) and with planning permission (discussed in Chapter 7). Of relevance too is the fact that France generates substantially more electricity than is consumed, and exports the balance. In addition, some 90 per cent of generation is already GHG-free. These factors reduce the impact of the drivers for wind power identified in relation to neighbouring countries.

UK

Despite some of the best wind and marine conditions in Europe and promising early initiatives,[15] the UK has not taken a pioneering role in renewables. Unlike France, Norway and Austria, the UK has little hydroelectric power. In the 1980s, progress in wind and marine power was blocked by the Conservative government's preference for nuclear and the exploitation of North Sea oil and gas. The non-fossil fuel obligation (NFFO) was established in the 1990s, consisting of a series of rounds of calls to tender. Whilst offering limited support to renewables, a key aim was to prop up nuclear power.[16] Because of these adverse framing conditions, RES-E stood at 1.7 per cent of supply in 1997.

The Labour party declared itself in favour of renewables development. Collier (1997: 105) noted that its 1994 programme set RES-E targets of 10 per cent by 2010, and 25 per cent by 2025. Although the Blair government came to power in 1997, no major legislation on renewables occurred till the Renewables Obligation of 2002. However, capital grants to support R&D and investment in renewables were made available. Policy drivers contained in the 2003 Energy White Paper were (1) to attain a 60 per cent reduction of CO_2 emissions by 2050, (2) maintain energy security, (3) promote competitive markets and (4) ensure that homes are heated affordably and adequately (Secretary of State for Trade and Industry, 2003: 11).

Policy design

The NFFO calls to tender system produced disappointing outcomes yet set counter-productive patterns which have persisted to the present. Competitive pressures to make low bids led commercial developers to search for economies of scale and to the 'wind rush' phenomenon of large wind farms being proposed for the windiest sites (Infield, 1995: 187). These were often in upland areas, in landscapes valued for their unspoilt beauty. Yet less than half of successful bids resulted in a comprehensive planning application (Toke, 2005a: 51). Where developers went ahead, they often simply announced their intentions to local communities, leading to antagonism and the creation of 'antiwind' protest groups (discussed in Chapter 8). The irregular sequence of calls to tender and limited deployment were inadequate to build up a domestic turbine industry, despite promising British ventures in the 1980s. By 2000, the UK had 400 MW of wind power, corresponding to a fifth of Spanish capacity and a tenth of German capacity.

A catalyst for renewal was European directive 2001/77/EC, containing a UK RES-E target of 10 per cent for 2010. Although experience with feed-in tariffs gained by the market leaders – Denmark, Germany and Spain – led to emulation in France, the UK government decided instead to use an untried policy instrument. The Renewables Obligation (RO) was established for England and Wales in April 2002, followed by equivalent instruments in Scotland and Northern Ireland. The basic principle is that RES-E generators sell their electricity by the usual means but they also receive a subsidy through the RO, which is financed by consumers through their bills. The RO is a form of 'Renewables Portfolio Standard' (RPS) which places an obligation on all electricity suppliers to derive specified proportions of electricity from renewables. The quota for 2002–3 was 3 per cent, rising in annual increments to 10.4 per cent for 2010–11, which matches the indicative target in the Renewables Directive. To reinforce industry confidence, in December 2003 the annual increment mechanism was extended, with 15.4 per cent as the target for 2015–16. To provide investment security, the scheme is guaranteed to run for 25 years, terminating on 31 March 2027.

Ofgem, the UK regulator, provides accredited RES-E generators with one Renewables Obligation Certificate (ROC) for each megawatt of their electricity. ROCs are a form of tradable 'green certificates'. Suppliers can either present sufficient ROCs to cover their obligation, or make up the balance by 'buying out' missing ROCs. Payments *via* the latter go into the 'buy out' fund. A 'buy out' price fixes the ceiling value of ROCs, and is set on an interventionist basis in the same way as is a feed-in tariff. In 2002, it was set at £30/MWh (circa 4.2 c€/kWh), but is adjusted *upwards* annually in line with the retail price index (unlike the German REFIT which contains a *downwards* mechanism). For 2006–7, it stood at £33.24 (Ofgem, 2006b). Revenue from the 'buy-out' fund is recycled to suppliers in proportion to their correctly presented ROCs. Thus the RO subsidy provides two income streams: the sale of ROCs *per se plus* an associated pro rata refund from the 'buy-out' fund.[17] In 2004–5, the 'buy-out' fund totalled over £135 million for England and Wales and over £17 million for Scotland, paying suppliers £13.66 for each of their ROCs and £19.99 for each RO (Scotland) certificate (Ofgem, 2006a: 3). Thus in 2004–5, the average worth of a ROC plus its associated 'buy out' refund was £45.05 (circa 6.3 c€/kWh), whilst an RO (Scotland) certificate was worth £51 (circa 7.1 c€/kWh) (Ofgem, 2006a: 4). But these figures are not fixed: they oscillate due to market movements. Due to the RO, suppliers can receive – or pay – significant amounts. In 2004–5, British Energy paid nearly £35 million whilst British Gas Trading Limited, Scottish and Southern Energy Supply and Powergen each received over £20 million (Ofgem, 2006a: 8 and Appendix 3). These large financial movements are a major incentive to investment in renewables, since in the UK the subsidy *alone* can be higher than some of the values for continental REFITs quoted above. It should be recalled that for UK RES-E generators, revenue from electricity sales is independent of the subsidy and is paid on top. The wholesale price of electricity has fluctuated considerably in the 2000s from £17/MWh hour to £70/MWh when gas prices peaked, with £45/MWh a typical figure in late 2006. In the UK, the total worth of a megawatt hour of RES-E is therefore the combination of two 'floating' values, that of ROCs and that of electricity.

In consequence, instrument design has created substantial investor risk. Uncertainties regarding the size of the 'buy-out' fund cause fluctuations in the market value of ROCs. In continental REFIT systems the burden of providing balancing services for wind power (discussed in Chapter 6) is passed to the TSO, but in the UK is financed by the generator. Further, the RO *excludes* the two corner-stones of continental systems which assure their stability, namely guaranteed prices and a purchase obligation. These design features lead to what Mitchell, Bauknecht and Connor (2006) called balancing risk, volume risk and price risk. To contain these risks, RES-E generators often enter into power purchase agreements with suppliers to ensure an outlet for their generation at agreed prices. But these risk management strategies are usually the preserve of large consortia. For small investors, the considerable market uncertainties created by the RO's design act as barriers to entry (unlike REFITs which encourage their participation).

In several continental systems, targets (for capacity and/or generation) are set for individual technologies, often within an indicative planning framework, ensuring clarity over goals and means. In the UK this approach was also rejected, since a core neo-liberal principle of the Blair government was to avoid prescription of the energy mix. The policy instrument supposedly leaves it to the market to decide on the best technologies to take forward. In practice, however, the RO is not 'technology blind' as claimed, since it fails to support emergent technologies whilst exercising positive discrimination in favour of near-market technologies. Indeed, the Carbon Trust (2006: 2) observed that the RO was 'designed to "pull through" lowest cost technologies sequentially'.

Given the high levels of subsidy, it is unsurprising that the market response has been enthusiastic. With an increase of 446 MW in 2005, aggregate wind power capacity in the UK reached 1380 MW. By the end of 2005 a further 4248 MW had been approved (IEA, 2006a: 250).

Policy discussion

At its launch, the RO was a new and untested variety of support instrument, but it was understood that it would increase consumer bills. The DTI estimated in 2001 that the increase would run at around £1 billion per annum at 2002 prices by 2010 – equivalent to a price increase of 5.7 per cent (National Audit Office, 2005: 12), whilst Eurelectric (2004: 104) calculated the total cost between 2002–3 and 2010–11 as £8.7 billion. Over 2002–5, its performance in terms of bringing new capacity on line was considerably lower than that of the feed-in tariffs in Germany and Spain, but slightly better than in France. Analysis has shown that substantial policy revisions were undertaken in continental countries. Yet nothing comparable occurred in the UK. Although a consultation on the RO was organised between March and June 2005, its outcomes left the RO's substantive features and settings unchanged.[18] In 2005–6 a wider UK 'Energy Review' was conducted, offering a forum for debate on energy options and raising the prospect of policy revision, with reforms expected in 2007.

Serious criticisms have been raised regarding the policy design of the RO. The House of Commons Environmental Audit Committee (2006: 50) stated that:

> the Renewables Obligation is an inflexible and inefficient mechanism for bringing to market a range of different renewable technologies due to the single flat rate incentive it provides. It effectively 'picks winners' by rewarding only the cheapest renewable technologies – mainly landfill gas and on-shore wind – and offers little or nothing to bring to market more expensive technologies such as offshore wind and marine. Such a mechanism contrasts sharply with the feed-in tariffs which have proved so effective in Spain and Germany in incentivising a range of different technologies.

The consequence of the inflexible design of the RO is that government cannot 'steer' policy towards targets as precisely as with REFITs. This is because REFITs allow sequential recalibration of settings, whereas the RO has in-built constraints to incremental improvements of the kind identified in the continental case-studies. These issues will be discussed in detail in Chapter 5.

Conclusions

The national case-studies have revealed that alongside similarities in policy support to wind power lie major differences. Policy *goals* are superficially comparable, but on closer inspection the relative weighting given to economic drivers (security of supply and national 'independence') versus environmental considerations varies. France and especially Spain are much concerned by the former, Denmark and Germany stress the latter, whilst the UK wavers between the two. Policy *specifications* have reflected national traditions, with state-initiated indicative planning playing an important guiding role in renewables deployment on the Continent, whereas the British preference for leaving the market to find solutions has led to a lack of clarity over targets and paths to attaining them. As regards choice of *instrument type*, in all cases the major instrument has been production subsidies, with investment grants being an accompanying measure. However, the characteristics of the instruments in use have not conformed to the stereotypes. Whereas REFITs are often assumed to be state-centred, practice shows that they have been used in a market-oriented manner. They can track wholesale electricity prices as in the new Danish system and especially the Spanish 'market option'. They can also respond to cost evolution in capital goods markets through the 'degression' mechanisms implemented in Germany and France. The frequent association between REFITs and 'fixed' prices is therefore a misunderstanding: price formation can be highly responsive to market movements. This is because REFITs give policy-makers considerable control over *instrument settings,* with periodic reviews allowing adjustments to match evolving circumstances and aims. On the other hand, the supposedly market-compatible RO has proved resistant to fine-tuning of its settings and to policy review because of its inflexible design. The *flexibility* and *versatility* demonstrated by feed-in tariffs are a major – and often overlooked – explanation of their success.

5
Drawing Policy Lessons from Cross-National Comparisons

Introduction

Policies to support wind power in Denmark, Germany and Spain have been developed over two decades, combining elements of continuity with systematic review and improvement. In France and the UK, policy development spans more than a decade, but the failed policy of calls to tender of the 1990s led to a switch in policy instruments in the early 2000s. This spread of European experience is a rich source of policy lessons. Accordingly, this chapter will focus on cross-national comparison and evaluation, and addresses the following core questions: Are policies on track to meet targets? What are the main lessons to be learnt from policy experiments across Europe? What recommendations can be made for improved policy design? Can a single policy template be applied across a number of nations? The first section analyses key consequences of the policy design of the two main categories of instruments currently used, namely feed-in tariffs and quota systems. The second section compares and evaluates these instruments, using the criteria of effectiveness and efficiency. The third section identifies explanations for differential performance, and on this basis recommendations for improved practice are made in the fourth section. The main argument is that on balance feed-in tariffs have performed better than quota systems, but that the lessons learnt from the former can – to an extent – be applied to the latter.

The consequences of policy design

The national case-studies in Chapter 4 demonstrated that the major instruments for the promotion of wind power have been REFIT and

RPS schemes. The merits and demerits of each have been discussed extensively in the literature, with analysts focusing on effectiveness and efficiency (Elliott, 2005; Haas *et al.*, 2004; Lauber, 2004, 2005b; Reiche and Bechberger, 2004; Ringel, 2006), stakeholder interests and investor groups (Enzensberger, Wietschel, and Rentz, 2002), security of investment versus risk (Mitchell, Bauknecht and Connor, 2006), market compatibility and the European liberalisation process (Meyer, 2003; Midttun and Koefoed, 2003), and the political dimensions of policy choice (Lauber, 2002; Hvelplund, 2005). These key themes will be explored further, with particular attention to the consequences of policy design for capacity increases, prices and categories of actor.

Capacity planning

An important policy issue is deciding on what quantity of renewables capacity is desirable and planning for its implementation, yet it rarely receives explicit commentary. Explanations from policy-makers for their choices of target are usually marked by their absence, with the Renewables Directive and the Renewables Obligation (RO) being totally silent on this point. Instead, broad policy drivers are invoked. But it is one thing to argue that renewables are desirable for energy security and environmental reasons, and another to specify a quantity by a particular date. Renewables enthusiasts have their own take on the question of quantity: they advocate that renewables should totally replace conventional energy. Their justifications are based on two dimensions of sustainability – the finite nature of conventional sources, with exhaustion being only a matter of time, and the finite nature of environmental sinks, notably the capacity of the atmosphere to absorb emissions. However, this normative vision does not address the empirical questions posed by the *sustainability transition*, particularly the availability of renewable energy sources, the viability and accessibility of conversion technologies, the need to anticipate the strategies of market actors (both incumbents and new entrants) and re-regulate to prevent blockages arising from oligopolies and market domination. On the other hand, policy-makers must grapple with these questions.

REFIT and RPS schemes provide particular answers to the question of capacity planning. In appearance, REFITs encourage an endless addition of capacity (although limits always arise in reality). Unsurprisingly, they are favoured by renewables enthusiasts. On the other hand, RPS schemes cap market share in relation to a time period. Consequently they are often favoured by incumbents, who consider that quotas allow them micro-manage new competitive pressures. But how is the 'right' level of generating capacity to be determined and by whom? Settling this matter has always been problematic because (a) the electricity system requires real-time balancing of rapidly changing demand and supply, and (b) peak demand vastly exceeds average load. Thus national ESIs have required an 'optimal' level of over-capacity.

Under the planning systems prevalent in the post-war period, fore-casters would extrapolate future economic growth rates from past per-formance and factor in a generous rise in generation capacity.[1] Even in Western market-based economies, these predictions would often trans-late directly into investment by publicly-owned utilities. However, inadequacies in forecasting, coupled with the shock of recession in the 1970s and early 1980s, led to misallocation of resources and excess capacity. Today, the dilemmas of forecasting are exacerbated by a variety of factors, of which two will be stressed. The first is the liberal-isation of European electricity markets. Liberalisation increases the opportunities for exports of electricity, meaning that capacity which is surplus to national consumption can – in principle – generate revenue streams from abroad. The second is the introduction into the grid of renewables. Because hydro and wind are weather dependent, they are subject to very large fluctuations in availability of the resource, with 'good' and 'bad' years arising in unpredictable sequence. This makes it difficult to anticipate the total quantity of *generation* from a specified *capacity* and complicates the planning issue of how much capacity to install. Additionally, because wind and photo-voltaic are randomly intermittent sources of generation, exact predictions for a quantity of generation at a point in time are impossible. This feature pulls into the equation the operational issues of grid management and system bal-ancing, but also has consequences for capacity planning . These factors make it problematic to determine the 'right' level of capacity and favour over-capacity. But at what point does a prudent level of over-capacity in the electricity system become 'excess capacity' and a waste of resources?

The question of deciding on an optimal level of over-capacity is a challenge to the neo-liberal paradigm which holds that the market is the correct means to allocate resources. This is because efficient markets are not normally characterised by endemic over-capacity, a sit-uation that companies normally seek to avoid. Regulation is therefore required to preserve 'optimal over-capacity' in the atypical electricity sector in order to avoid market failure, diversify sourcing and ensure adequate supply. In the post-liberalisation context, the expansion of RES-E is an apposite example of this regulatory response. Yet intervention in the promotion of renewables runs counter to the liberalisation agenda and represents a return to planning by stealth. The result is a 'halfway house' between state and market.

The uneasy compromise between state and market is reflected in the 'one-sided' nature of REFITs and RPS. With each type of production subsidy, the policy-maker intervenes in one dimension of the policy equation – namely, prices *or* quantities – whilst apparently leaving the other dimension up to the market. Thus REFITs control prices (by setting a regulated tariff per kilowatt hour), but in appearance leave the market to decide on quantity (namely the level of capacity to install). On the other hand, RPS schemes control quantities, whilst apparently leaving prices to the market. However, appearances prove deceptive. In practice, each scheme must deal with prices *and* capacities, due to the necessary interplay between them. With REFITs, each price setting creates its own level of market pull, since higher prices incentivise greater capacity. In quota schemes, the price of 'green certificates' is determined within an artificial market created by intervention and whose parameters are set by policy-makers. Although in each case government appears to be setting only one term in the equation, the other term is also strongly influenced. Indeed, among the nations utilising REFITs, capacity targets were specified at various junctures in Denmark, Spain and France. However, the interaction between prices and capacities has not always been well-understood, as it is a product of an ongoing 'learning by doing' process. Thus its consequences have rarely been acknowledged or acted upon. Moreover, the market compatibility of RPS schemes has been exaggerated by understating the reality of government intervention in policy design and instrument settings. In summary, each instrument contain elements of *both* price control and capacity planning, but further policy learning is required to coordinate management of the two dimensions.

Small-scale or large-scale capitalism?

The choice of policy instrument has consequences for patterns of renewables ownership. REFITs can fit with either small-scale or large-scale capitalism. In Denmark and Germany, the wind sector is characterised by the small-scale capitalism of individual and cooperative investment. Small investors often have a longer-term perspective on renewables holdings than large firms and unit trust funds, where ownership 'churn' is frequent. Turbine ownership is sometimes compared in Denmark to a pension plan, clocking up income before and during an individual's retirement. Attracting private investors to wind power has particular consequences. It increases societal engagement in alternative and sustainable technologies, and fosters social acceptance. It also reduces the scale of policy support that is required since, as noted by Enzensberger, Wietschel and Rentz (2002: 799) 'small private investors tend to be more risk averse than professional players. However, their profitability expectations are also usually lower'. Thus prices and revenues can be set lower, provided they are guaranteed. On that basis, REFITs in Denmark and Germany successfully introduced new private entrants into the ESI and reduced the market dominance of the incumbents.

Although Danish and German utilities showed disinterest and even antagonism towards the emergent wind sector in their domestic markets, Spanish utilities overcome their suspicion of wind power relatively early, indicating that REFITs can appeal to large-scale capital too. Further, Spanish policy-makers were able to shape REFITs into market-oriented variants because of the involvement of large corporations who were accustomed to market schemes. The French policy model of 2001 appeared to be steering towards small-scale ownership, due to the 12 MW ceiling on eligibility for the REFIT. Yet calls to tender for larger wind farms gave opportunities to corporations. In one sense, this configuration allowed for evolutionary selection in development pathways. But in another it gave an ambiguous response to the question of ownership. The appeal to small investors remained underdeveloped, partly due to the limited availability of investment vehicles in France, but partly due to timing. France came late to the sector and lacked the scope for a 20-year learning curve by small investors as seen in Denmark and Germany. By the 2000s, the upscaling of turbines and installations necessitated greater financial reserves than in the sector's early days. By default, the development model in France became one of medium to large wind farms, financed by business consortia. The abolition of the 12 MW ceiling in 2006 consecrated the model of large-scale capitalism in the wind sector.

By contrast, RPS schemes such as the RO are attractive almost exclusively to large enterprises. New capacity has been brought on-line by utilities and specialised developers (who often sell on their projects to the utilities and large firms). Meanwhile, the non-corporate sector – farmers, cooperatives and citizen investment initiatives – that is so important in Germany and Denmark barely exists in the UK.[2] As identified in Chapter 4, the price risks and market uncertainties inherent to the RO drive out all but the most courageous of small investors. Noting that the BWEA was particularly supportive of an RPS in the UK,

Lauber (2004: 1412) commented that 'the technological culture RPS schemes are likely to produce does not differ from that of the utilities, based as it is on large-scale, centralised production of RES-E to achieve commercial advantage, and indeed utilities are viewed as essential actors'. A preference for market-based systems is also the position of the European electricity industry.[3] The structural consequence of the RO is that the UK wind sector is isomorphous with the electricity supply industry, being characterised by large-scale capital looking for short-term profit. Thus in response to the question of ownership, the RO produced a context-specific and unambiguous response. If small-scale development by 'gifted amateurs' was not viable in the short run-up to 2010, then the answer was big investors and the utilities. But in choosing the RO over a REFIT, few new competitors emerged. Thus whereas Danish and German REFITs challenged the position of market incumbents, the RO reinforced it. It is not surprising then that in Germany REFITs met legal challenges from the utilities, but in the UK the utilities were favourable to the RO. The irony is that, because of a wave of international takeovers, utilities such as E.ON and RWE investing in wind power in the UK were the same companies that had shown resistance to it in their domestic markets. The corporate takeover of alternative energy on the back of a subsidy scheme provides an instance of how 'the empire strikes back'.

However the UK's choice of an RPS was not inevitable. By the 2000s the Spanish REFIT, with its 'market option', had also developed on the basis of industry isomorphism – namely large-scale capital and utility involvement in the wind sector – and was a feasible solution to the ownership/investment problem as posed in the UK. This is important, since the implications of selecting RPS over REFITs in terms of effectiveness and efficiency are considerable.

Comparative evaluation of policy instruments

Continental REFITs and the British RO will next be compared and evaluated using the criteria of effectiveness and efficiency. The criterion of effectiveness concerns the quantity of new capacity coming on line and the timeliness of build in relation to targets. The criterion of efficiency relates primarily to the price competitiveness of generation. But it also concerns other dimensions of competition, notably equipment costs. The ideal system would combine high effectiveness with high efficiency. But is there a tension between the two, with the entailment that one must be traded off against the other? Given that effectiveness is measured in terms of capacity growth, it is only logical that systems which have no cap on build (namely REFITs) can perform better on this criterion than those that do (RPS). And one might expect that the cost of fast growth would be higher prices (and lower efficiency). Indeed this criticism was frequently made in the 1990s. But outcomes have turned out to be more complex than this.

Effectiveness of policy instruments

There is a clear correlation between capacity growth and the use of feed-in tariffs. In January 2006, the market leaders Germany, Spain and Denmark, with 18,427 MW, 10,028 MW and 3127 MW respectively, together accounted for 72 per cent of the 43,604 MW of capacity in Europe. With the UK on 1342 MW, the effectiveness of the RO is lower at face value. However, the scheme has been in place for considerably less time. The French feed-in tariff – which is of comparable vintage to the RO – led to a still lower rate of build, at 770 MW. Thus whilst the headline capacity figures are suggestive, they cannot tell the whole story.

A second mode of assessment is to track progress in relation to targets. In Table 5.1, the first column lists the reference countries whilst the second reproduces their market share targets as contained in the European Renewables Directive. The problem is that percentages have little meaning until converted into terawatt hours of generation. Accordingly, the third column presents estimates of gross electricity

Table 5.1   Extrapolation of RES-E targets contained in the European Renewables Directive

Country	2010 targets %	Projected total consumption in 2010 TWh	Projected 2010 RES-E generation targets TWh	RES-E in 2004 TWh	RES-E in 1997 TWh	Extra RES-E generation relative to 1997 to achieve 2010 targets TWh
Denmark	29	37.9	10.99	10.25	3.21	7.78
France	21	532	111.72	71.17	66	45.72
Germany	12.5	545	68.13	68.48	24.91	43.22
Spain	29.4	306	89.96	56.73	37.15	52.81
UK	10	434.8	43.48	15.93	7.04	36.34

Sources: Official Journal of the European Communities, 27.10.2001, L283/39
    Eurelectric (2004: 14)
    IEA (2005b: I.36)

consumption in 2010, as produced by Eurelectric (2004: 14). On this basis, column four converts RES-E market share targets into generation targets. Column five gives 2004 outcomes, based on IEA (2005b) data which aggregate hydro, solar, wind, biomass and waste. This gives a measure of progress to 2010 targets. Column six is a reminder of RES-E generation in 1997. In relation to this 1997 baseline, the final column estimates the total quantity of extra renewables output required to meet the estimated 2010 generation target.

These data suggest that Germany and Denmark are set to attain or even overshoot their targets. This is a further indication of the effectiveness of their REFIT policies. Spain has an ambitious renewables target and is poised for *capacity* expansion, but success is not assured in meeting *generation* targets. The 2010 targets for wind power are challenging: 20,155 MW of capacity and 45,111 GWh of generation, involving investments of over €23 billion (IDAE, 2006: 7 and 14). This will place wind power 'among the large-scale generation technologies' (AEE, 2006: 13). But it assumes very high build rates of around 2000 MW p.a., as compared to an average build rate of 1600 MW p.a. over 2002–4. It also assumes higher generation per MW from new installations than achieved in 2004–5 for current wind farms, yet new build will be forced mostly onto lower wind speed sites. If Spain is trying to square the circle, it is because increases in output from renewables other than wind have been modest over the past decade. Indeed, hydro has seen generation short-falls due to drought.

France too has seen a reduction in output from hydro, but with little growth in any other renewable source. Wind power has displayed disappointing performance. Given that the current average capacity factor is lower than anticipated at a national average of 1798 full load hours or 20 per cent (Bataille and Birraux, 2006: 131), the required capacity from wind power to meet the 2010 target would need to be at least 10,000 MW, and probably more.[4] In 2006, France crossed the 1000 MW threshold. In 2005 several thousand megawatts of capacity were in the pipeline, with 1557 MW given planning permission and a further 3198 MW in process (Ministère de l'Economie, des Finances et de l'Industrie, 2005b). This suggests that considerable new capacity will come on-line later this decade. But France has a very ambitious target which, if achieved, would make it by far the largest RES-E producer in Europe. On current trends, however, the 2010 target will not be met.

The UK too is some way short of meeting targets.[5] A complication in analysing the British case arises from the lack of official capacity and generation targets for 2010. As a consequence of the principle of 'technology neutrality', targets are not set for individual technologies, creating uncertainties over the future sourcing mix. This makes it difficult to identify whether there is a short-fall in the wind power component. In 2004–5, the main RES-E sources were landfill gas (30 per cent), co-firing (19 per cent) hydro (18 per cent) and wind power (16 per cent) (Ofgem, 2006a: 18). But this order is expected to be reversed, with wind power becoming the leading source. On October 25th 2004, Mike O'Brien (then Energy Minister) told the House of Commons that:

> The UK has set a demanding target for 10 per cent of our electricity generation to be supplied by renewable energy by 2010. (...) We expect 7 or 8 per cent of the 10 per cent generation to come from wind energy. Other technologies will be hard pushed to produce the rest. (...) Roughly half the 7 or 8 per cent wind comes from onshore wind generation and half from offshore wind generation.

Here too the percentages need to be converted into generation targets. 'Semi-official' calculations can be found in the literature. The Sustainable Development Commission (2005: 11) claimed that with an average capacity factor of 30 per cent, 9500 MW of wind power could produce 31,500 GWh of electricity, sufficient to put the UK on track for the 2010 target of 10.4 per cent. However, caveats regarding average capacity factors, demand increases and the availability of generation from other renewables are in order. Given the uncertainties, a higher figure of 10,000 MW plus of wind power capacity for 2010 may be more realistic. This would require over 2000 MW of new capacity each year. At the end of 2005, 4248 MW of additional capacity had been approved but not built (IEA, 2006a: 250). Assuming that most of these projects reach completion, cumulative capacity is set to grow dramatically over 2006–8. Indeed, the rate of build in the UK is faster than in France, indicating that REFITs are not automatically more effective than RPS. Nevertheless, a short-fall in relation to the 2010 target seems likely.

Why then are quotas not being fulfilled in the UK? Experience gained between 2001–5 has revealed that the design of the RO is flawed. In principle, the RO is governed by market forces yet policymakers sets the quota for RES-E and the 'buy-out' price for ROCs. Thus the system is far from forming a typical, nor a perfectly functioning, market. Once targets come with reach, the system presents *disincentives to new investors* because, as noted by the National Audit Office (2005:

18) 'as the supply of Renewable Obligation certificates increases relative to the obligation level, their price reduces'. This leads to the danger of a so-called 'cliff edge', beyond which the value of ROCs collapses due to over-supply. As explained by Oxera (2005: 18) 'the value of ROCs will fall towards the buy-out price as the total volume of renewable generation approaches the obligation size'. In consequence, current beneficiaries of ROCs may have a material preference for quotas *not* to be met. The consequence of disincentives to expansion for both recipients and potential entrants is that a cap on expansion is encountered some way short of any quota. The entailment forecast by the Carbon Trust (2006: 2) is that 'renewable energy penetration would be only three-quarters towards achieving the target for 2010, and only halfway towards achieving the 2020 aspiration for 20 per cent'. These would be serious limits on the effectiveness of the RO.

A shortcoming of the policy instrument is that, since ROCs only have value for suppliers, it gives opportunities to major players to control the market. Although several dozen suppliers exist in the UK, because of market concentration most of the obligation falls on very few. In England and Wales, over 50 per cent of the obligation in 2005 fell on three (E.ON UK, RWE npower and London Energy/EDF) and the biggest six accounted for 90 per cent; over 70 per cent of the Scottish RO was discharged by ScottishPower and SSE Energy Supply (Ofgem, 2006b: 5–6). The financial incentives of the RO have encouraged the utilities to own and operate RES-E portfolios. Although the 'independent' generators (who are sizeable companies) undertake some 60 per cent of RES-E generation, they receive only such value from ROCs as suppliers give them. In theory, 'independents' can auction their ROCs (and electricity) on spot markets, but this is not always attractive in practice because of low liquidity in the ROCs market and because they generally need power purchase agreements (PPAs) as collateral for financial-market loans. PPAs are negotiated between generators and suppliers in a context of unequal market power in which suppliers dominate. Evidence is emerging that suppliers may exploit their market power during negotiations. As noted by Oxera (2005: 19):

> suppliers may be able to capture a significant proportion of the ROC value for themselves. Furthermore, the high concentration of RO share among the six main suppliers means that they may have an incentive to restrict the total volume of renewable generation available in the market in order to maximise the total value of the ROCs that they control.

In consequence, only a proportion of the worth of ROCs is effectively passed on to developers, with the suppliers retaining the balance. This 'leakage' of subsidy may be justified by the context of the forward contract in that the supplier who provides the security of a 15-year contract takes on risks related to ROC values as well as the cost of providing balancing services, each of which merit remuneration. However, the danger of a restriction in the total volume of renewable generation gives pause for thought. A major design flaw of the RO is that the total cost to customers is the same, *regardless of how much renewables capacity is built*. It is this feature which raises concern over the inherent effectiveness of the policy instrument.

In summary, the countries closest to targets are those which have used REFITs over the long term, namely Denmark, Germany and Spain. The recent implementation of a REFIT in France has not encouraged investment on the same scale. Indeed, the rate of build in France is lower than in the UK, as is the quantity of projects which have received planning approval. Yet the RPS implemented in the UK has major flaws, with the consequence that quotas will probably not be fulfilled. However, choice and setting of the policy instrument *alone* do not explain differences in outcomes across the reference countries. Other explanatory variables are involved and these will be considered in this chapter and those that follow.

Efficiency of policy instruments

A key observation is that REFITs have proved cheaper than the RO. Illustrative data are presented in Table 5.2.

Table 5.2   **Prices of wind-generated electricity, 2004–5 (in €c/kWh)**

REFITs	
Germany	8.5 (6.5)
France	8.4
Portugal	7.5–7.9
Austria	7.8
Spain	6.3–7.5
Greece	6.4
RPS	
UK	10.1
Italy	15.5

Source: BWE (2005: 17)

Generation prices in renewables schemes

There are several reasons for lower generation prices under REFIT schemes. Predictable tariffs give a high measure of investor security. Low risks and guaranteed revenues translate into low interest rates on loans. Further, private investors are often willing to accept lower rates of return on investment than corporate investors. As we have seen, they are active in the REFIT countries but largely absent in the UK. In Denmark and Germany, RES-E generators do not have to pay for balancing services, whilst grid connection costs are relatively favourable. But in the UK, connection and balancing costs are negotiated between developers and utilities: part of the receipts from ROCs is used to pay those costs.

However, a further problem is that market-oriented support mechanisms have reflected inflationary pressures in conventional generation. A major cause of higher electricity prices in 2005–6 was the steep rise in gas prices. To the extent that wind power is a price taker, wholesale price inflation produces wind fall profits. Thus high RES-E prices in the UK arose from the addition of escalating wholesale electricity prices and high ROC values, due to short-falls against quota. According to Ofgem (2006a: 4), a ROC in 2004–5 was worth £45.05 on average to a supplier in England and Wales and £51.38 in Scotland.[6] Generators also sell their electricity, and over the 2002–5 period wholesale prices ranged between £17 and £25. Thus a representative price of UK RES-E in 2004–5 was of the order of £70 per MWh, roughly 10 €c/kWh.[7] But later in 2005 prices reached £90.50 per MWh in 2005 (IEA, 2006a: 254), roughly 14 €c/kWh. These levels are considerably higher than continental feed-in tariffs. They indicate the problems associated with the artificial market created for UK 'green certificates'. When wholesale prices move upward for an RO producer, *total* income from electricity sales plus ROCs tends to increase.

Yet as the wholesale market price of electricity *increases* the need for subsidy *reduces* – but there is currently no mechanism to achieve this. In consequence, the RO was making wind power progressively more expensive to UK consumers at a time when degressive feed-in rates were making it cheaper in Germany and France. This is a problematic outcome, especially when Britain's more abundant wind resource is factored in. Higher wind speeds make installations more profitable, and less needy of support. This combination of factors resulted in over-payment for wind power. The National Audit Office (2005: 40) concluded that 'the level of support provided by the obligation is greater than necessary to ensure that most new onshore wind and large landfill gas projects are developed', going on to quantify this excess support as being between a third and a half more that what is required. Similarly the Carbon Trust (2006: 11) found that 'on-shore wind is nearly cost-competitive today, with current cost estimates at around 5 p/kWh net of balancing costs (to deal with intermittency issues). This compares to a current wholesale price in the order of 4.5 p/kWh'. In relation to a spread of technologies, Ofgem (2007: 8) reported that 'at an average price of £45/MWh (close to the current wholesale price) all of the existing deployed technologies are economic without the need for any further support suggesting that nearly all of the RO subsidy is in excess'. These analyses by the major institutional actors all point to the efficiency of the RO being unacceptably low.

Price convergence is changing the basis of need for subsidy. Indexation to wholesale prices is a contributory cause of the overpayment problem. In Spain, wind fall gains arose in 2006 because feed-in premiums track market prices. However, the risk of over-paying existing capacity contributed to a revision of Spanish payment mechanisms in early 2007. To avoid the indexation problem, Denmark has capped total remuneration (electricity sales plus premium). The problem of over-support is dealt with in Germany and France by the principle of 'degression'. Not only does policy support reduce over time, but regulated wind power tariffs can dampen average wholesale price increases. In the current context of rising fuel costs, German consumers are partly cushioned because regulated wind power prices have been falling whereas market prices from conventional generation have been rising. Over time, convergence between the two prices curves can mean that consumers receive a measure of compensation for subsidies. But in the market-driven support systems of Spain and the UK conventional fuel inflation increased the profitability of RES-E in 2006 and increased burdens on customers. Spanish policy-makers moved quickly to correct this loss of efficiency, but there was no rapid reaction in the UK.

Optimal efficiency is achieved where remuneration systems are 'cost reflective'. A 'cost reflective' approach sets subsidies at levels commensurate with investment costs and adequate return on capital, but without over-burdening taxpayers or customers. This achieves what Chabot (2001) called 'fair and efficient' payments. In practice, this has proved easier to attain within continental REFITs than the UK RO. The claim that the RO is a market-oriented instrument which encourages price competition is contradicted by windfall profits due to electricity price inflation and the consequent failure to be 'cost reflective'. In the words of the regulator, OFGEM, a scheme which will cost £32 billion over its lifetime offers 'poor value for money for customers' (Ofgem, 2007: 1), when 'cheaper and simpler' options are available.

Competition in wind power equipment markets

Regulated feed-in tariffs can reflect investment costs more accurately. To understand why requires closer examination of equipment markets. As shown by Hvelplund (2001b: 182–3) and Meyer (2003: 674), the nature of competition in the wind power sector (as well as PV and hydro) is fundamentally different to conventional generation. In the fossil fuel sector, competition arises from capital inputs, labour inputs and especially fuel costs. But wind installations are different. Over 80 per cent of the cost of wind power is in capital expenditure,[8] with the remainder going to ground rent, operation and maintenance, and so forth.[9] Post-construction labour costs are thus low, whilst 'fuel' is free. Wind turbines have been likened to energy robots,[10] whose 'competitiveness' is a function of nature in that output is governed primarily by how windy it is. The difference between 'good' and 'bad' wind years can lead to a 30 per cent variation in output. These atypical features do not correspond to normal market competition, yet renewables policy design must respond to them.

A more typical form of competition occurs in relation to equipment supply, namely the turbine and its components (the generator, gearbox, blades, tower and associated infrastructural hardware). In the past, REFITs have encouraged competition between turbine manufacturers. Since the price received by generators was regulated and the same for all, investors would go to 'best value' equipment suppliers. This stimulated technical improvements and lower prices, both through conventional competition in price formation between firms and from economies of scale in expanding markets. The degression principle used in Germany and France sought to encourage these effects.

However, the international business environment for equipment supply altered in the mid-2000s. In 2005–6, turbines prices rose. The major cause was increased world prices for raw materials, particularly steel. But turbines manufacturers were in any case seeking to increase prices, having gone through a period of losses.[11] The opportunity to do so was presented by excess demand in world markets. In Europe and North America, public policies to promote wind power resulted in a situation where demand exceeded supply. The setting of arbitrary deadlines – such as 2010 by the Renewables Directive – contributed to this effect. With turbines in shortage, manufacturers could sell into those markets that pay best. This placed the shoe on the other foot. Competitive pressures on turbine suppliers have eased, whilst national support schemes now compete to attract suppliers for products and inward investment in new manufacturing facilities.

The situation is complicated by the national origins of the wind turbine industry. German and Spanish suppliers have incentives to service home markets first. But with next to no home market, the Danes must export. On the other hand, countries such as the UK and France have very little domestic supply and are dependent on imports. In theory, excess demand would encourage new entrants into the market. This is happening in Asia, notably, with Suzlon and Goldwind. In Europe, however, a market shake-up has led to turbine manufacturers and their suppliers disappearing, merging or being taken over by large conglomerates. The resulting oligopoly, backed by its engineers and patent lawyers, has the technological, financial and legal means to maintain its stranglehold on the market, at least for the medium term.

In this context, a mechanism such as the RO is unable to encourage market competition. On the contrary, turbine shortages contribute to restrictions in RES-E supply in the UK, and so help keep ROC values high. On the other hand, continental REFITs have a track record of encouraging competition in equipment supply. Even in a context of equipment shortages, degression mechanisms cap the total price that users must pay, and constitute a form of consumer protection.

In summary, regulated tariffs – especially where a degression principle is applied – have the advantage of adhering more closely to the commercial reality of constructing and operating wind turbines. The 'upfront' investment costs are high, but post-build costs are low. Interest on commercial loans is often higher in the initial investment and running-in period, and lower thereafter. A structure of higher initial policy support, which gradually tapers to nil over an appropriate 'pay back' period, reflects the investment profile. This produces an efficiency gain for customers, and represents a better channelling of subsidies. In a phrase, policy mechanisms need to 'get the prices right'. Despite their imperfections, REFITs have enjoyed a greater measure of success in doing so.

Explanations of differential performance

The core explanation for the performance differentials between REFITs and the RO relates to the flexibility, adaptability and 'steerability' of each. In order for policy-makers to reach policy *goals*, changes are periodically required in the *settings* of policy instruments. This process can be likened to steering a vehicle. Adjustment of the controls of a vehicle in response to real time developments ensures safe arrival at the chosen destination. The continental case-studies have demonstrated that feed-in tariffs give the policy-maker flexibility in the setting of a wide range of controls, namely:

- level (the tariff itself);
- destination (the conversion technologies targeted);
- volume (capping of subsidy in relation to output and/or to wholesale prices);
- time (period of eligibility);
- scale (size of installation);
- location (site conditions such as wind speed);
- category (all new build, just 'repowering', or other);
- recipient (small investor or large utility).

Although no single system has used all these parameters, each can be in found in one or more of the systems surveyed.

Continental experience has demonstrated that evolving circumstances require periodic adjustments to instrument settings, since no choice is 'right' forever. REFIT systems have seen repeated adjustments in parameters in order to correct undershoot or overshoot. Whilst various settings have proved amenable to fine-tuning, clearly the major variable is the tariff level which typically has been *reduced*. This demonstrates that policy amendment is possible without significant 'regulatory risk'. On the other hand, the UK system has so far not been recalibrated. Indeed, the RO is characterised by having few recalibration mechanisms. The key settings are the quota, the setting of the buy-out price and the period of eligibility. In effect, these were set once and for all during 2002–3, leaving everything else to the 'market'. At face value, UK policy-makers can do little till 2027 regardless of whether the initial parameter settings were 'right'. Some choice do remain, however. They can increase quotas for the post-2015 period and extend the time-frame, or marginally alter destination of subsidies.[12] Grid connection issues are subject to regulation.[13] But this is an impoverished range of choices as compared to REFITs. In consequence, under current arrangements installations built in the 2000s and maintained in good repair are entitled to subsidy through till 2027, regardless of 'pay-back' periods, current and future profitability levels and whether they actually need subsidy. A further question mark hangs over new build in the post-2015 period. Even if the quota were increased yearly on the current pattern, new build would be disadvantaged by the short remaining term to 2027 and consequently may not be incentivised. In this scenario, the perverse result of the lifting of post-2015 quota could be to direct wind fall profits to existing installations – an outcome which clearly must be avoided.[14] But with so few parameters available for setting, the RO lacks an effective 'steering-wheel'. This lack of 'steerability' needs to be corrected, and a more equitable and cost reflective system instituted.

Turning then to a summary of the main policy lessons, experience has shown that the advantages of 'feed-in tariffs' in relation to wind power are:

- guaranteed and predictable prices together with a power purchase obligation create a low risk investment environment;
- the instrument thereby encourages a wide range of new entrants and dependence on utilities is decreased;
- a cost reflective approach is achieved by differentiation between technologies;
- degressive rates stimulate competition between manufacturers, and reduce costs for customers;
- expansion of the wind turbine manufacturing sector is stimulated;
- no implications for the government budget, as subsidies are paid by customers.

The disadvantages displayed by the RO are that:

- relatively high risks, due to long-term uncertainties over the size of the buy out fund, which only large consortia will shoulder in return for a substantial risk premium;
- the lack of obligatory purchase increases the market dominance of the utilities by making RES-E generators dependent on suppliers and further reducing the scope for new entrants;
- the lack of differentiation between technologies produces a pricing system which is incapable of being cost reflective;
- the capacity of the RO to foster either industrial development or market competition has not been demonstrated.
- policy-makers are failing to 'steer' the policy regime due to the lack of clear goals and the inflexibility of the policy instrument.

Like REFITs, the RO has the advantage of presenting no implications for the government budget. A further advantage is that the RO provides a route to the market integration of renewables by developing commercial relationships between actors in the ESI. On the other hand, REFITs are being 're-engineered' in various ways to improve market integration, notably in Spain and Denmark (but not as yet in Germany). This theme will be taken up in the next chapter. Having summarised core lessons, we next turn to recommendations.

Recommendations on the development of policy instruments

The debate over the relative merits of 'feed-in tariffs' and 'renewables portfolio standards' is far from over. At the EU level, lessons have been drawn from national policy experiences in RES-E support. In the late 1990s, RPS policies were in the ascendant within the European Commission, since they were considered more compatible with the liberalisation agenda. However, the latest EU progress report on the Renewables Directive found that REFITS were more effective *and* cheaper than quota systems (European Commission, 2005: 7). Successful RES-E deployment in Germany, Spain and Denmark, and the resurgence in the mid-2000s of security of supply arguments, swung the Commission back to accepting feed-in tariffs. At the level of theory, it is now widely recognised that each production subsidy scheme creates a 'political market'. In other words, they all arise from political intervention yet all have a market dimension. Their hybrid nature makes judgement of their relative degrees of 'market compatibility' hard to call,[15] so reducing the impact of ideological arguments drawn from neo-liberalism. At the level of practice, the Commission could hardly undo the successful policies of the market leaders. In consequence, the tide has turned in the process of regulatory competition. Having attained leadership through feed-in tariffs, Germany, Denmark and Spain have an interest in perpetuating and disseminating that system. For the moment, the latter is a step too far for the Commission, which has opted for the compromise position that it was premature to institute a harmonised European support system on either an RPS or REFIT basis.

At the national level, calls are still heard for a wholesale switch from one system to the other. In the UK, the Carbon Trust (2006: 24) concluded that 'the option of retaining the current policy in its present form is very costly', and so recommended a transition to a 'renewable development premium', a variety of feed-in tariff. In Germany the VDEW (2005) called for a switch to an RPS. However, governments become locked-in by previous policy decisions, due to the problems raised by regulatory risk, the opposition of key actors and the uncertain marginal benefits. The need to provide a stable investment environment for new RES-E technologies merits stress. The lower risks to investors in the German and Spanish systems contribute to their superior performance. An attempt to effect wholesale change can backfire, as happened in Denmark in 1999–2003 with the abandoned switch to an RPS and the collapse of the home market. Wholesale policy change does of course present risk, and where attempted requires considerable circumspection.

However, 'path dependence' constraints should not be an excuse to preserve the status quo indefinitely and, moreover, do not rule out incremental but substantial improvements. The longitudinal, cross-national comparison undertaken here indicates that an 'optimised' support system for RES-E would be based on the following principles:

- technology differentiation;
- precise and meaningful targets;
- cost reflective subsidies; and
- synergy of energy policy and industrial policy measures.

The core recommendation made here is that future policy explicitly seeks to combine *all* these principles into practice. However, whether this could be done with one instrument or would require multiple instruments merits new research.

Of these, cost reflective support is arguably the most important. However, this principle is impossible to implement without the others. Technology differentiation recognises the characteristics of renewable energy conversion technologies, and the relative maturity of each in terms of the quality and performance of products, available expertise and the development of manufacturing capability. Precise targets – specifying capacity and output – encourage openness, realism and dialogue with stakeholders. These require a context of indicative planning. In practice, subsidy need is a dependent variable arising from the *combination* of technology and target. Only once the combination has been defined is it possible to set a cost reflective level of subsidy. This is easier for conversion technologies which are close to market maturity, of which wind power is clearly one.

The costs of wind power investment are well-known and are relatively predictable over the long term, unlike nuclear power or gas-fired generation. Subsidies must reflect those costs, being set at a level which effectively encourages a chosen rate of growth, offers a fair and reliable return to investors but without leading to excess or windfall profits. To this end, subsidies should be capped in relation to output, costs and prices. This already occurs in the German, French and Danish contexts. Although the calculations vary in their technical details, the core principle is that aggregate income – whether from the feed-in tariff alone or the combination of electricity sales plus premium – does not exceed a defined level. Once that level is passed, the subsidy is withdrawn. Expressed as a general principle, the threshold level for capping is defined by the relationship between the average wholesale price of electricity (calculated yearly) and the real cost of RES-E. Where RES-E generation from a particular source costs more than the wholesale price, an 'environmental premium' is worth paying in specified circumstances. But where the wholesale price systematically exceeds RES-E generation costs (calculated to include a fair return on investment), subsidy should be stopped with rapid effect. It is recommended that this principle be applied both to the RO and to REFIT schemes. Applied to the RO, it avoids so-called 'deadweight' or payments to schemes in no need of subsidy. Applied to REFIT schemes, it encourages greater market integration by giving the option to sell outside of the 'fixed' tariff where wholesale prices are higher and more attractive.

An associated effect of setting subsidy caps in relation to specific conversion technologies and their actual costs is to increase the range of instrument settings. For example, this approach helps resolve locational issues. A single payment to all categories of sites (as is offered under the RO) is an incentive to seek out the windiest sites. Although this makes economic sense, it encourages the 'wind rush' phenomenon of concentration of build in particular areas, often of high landscape value. Setting a higher subsidy cap for installations in costlier, lower wind speed sites would mean that, over time, they would receive a higher subsidy than on high wind sites but the mechanism would encourage dispersion. However, a minimum productivity threshold for eligibility would also need to be set, as is done in Germany.

Turning to wider recommendations, new regulatory challenges arise from the development of the renewables sector and its claims for financial support. First and foremost, the payment of subsidies should, in exchange, entail transparency and accountability on the part of recipients. In a subsidised regime, data on costs, generation and profitability levels need to be reported to the authorities, whilst information on subsidies should be available to the public. This information should not be hidden behind the veil of 'confidentiality'. As demonstrated in the Danish case, openness encourages public confidence, stimulates market competition and helps improve the quality of decision-making. It is important that national energy regulators be enjoined with particular responsibilities and powers to achieve these purposes of transparency and accountability in relation to the substantial and increasing consumer subsidies to RES-E, and help achieve 'value for money' for customers in attaining policy goals.

In continental feed-in systems, the policy review process includes the setting of subsidy levels. This necessitates audit of their effectiveness and efficiency and so incorporates regulatory functions. With the British RO , this intensity of scrutiny does not occur. Yet 'leakage' of the value of ROCs to the supplier means that a proportion of consumer subsidies either fails to reach current generators or fails to incentivise further investment. This is unsatisfactory in a subsidy regime where the total cost to consumers is the same, regardless of the quantity of RES-E generation. Yet not enough is known about why and how such failures occurs, and to date it would appear that little has been done about it. Greater surveillance and reporting are required to improve transparency and efficiency. In all countries, audit systems focusing on transparent and fair distribution of the charges of grid connection, access and transmission charges and of the costs of balancing services would benefit both RES-E generators and customers.

Conclusions

Wind power in Europe has been a 'political market' in that capacity growth is causally linked to public policy – in general, the more supportive the policy, the bigger the expansion, and the more predictable and continuous the scheme is, the stronger the rate of expansion. Yet 'throwing money' at the problem is not enough, especially if it results in poor value. Policy learning over two decades has revealed substantial differences in levels of effectiveness and efficiency of the various support schemes. Lessons can be learnt from cross-national comparison in order to improve practice. Whilst harmonisation of RES-E support across the EU on the basis of a single instrument remains hazardous, common guiding principles for instrument design have been identified. A core requirement for the next generation of subsidy schemes is to systematically reflect price convergence between conventional and renewable sources, irrespective of whether the convergence is caused by rising conventional fuel costs, falling RES-E technology costs, or other factors such as emission trading schemes. Price convergence is changing the basis of need for subsidy. Support schemes will need to be sufficiently flexible to respond *quickly* to developments, to allocate subsidies where there is genuine need and valid purpose, and to withdraw them once the need disappears.

To date, REFITs have proved more successful in serving the *energy policy* purposes of diversified sourcing and increased generation *via* RES-E and the *industrial policy* purposes of promoting the manufacturing sector. In the market-leading countries, higher energy costs have been offset by technological leadership, employment creation and export opportunities in a new industry. Market leadership has the effect of compounding the benefits of an early choice of successful instruments. The problem for 'late entry' countries is that not all of these benefits can be realised by them. However, incremental improvements remain desirable in order to get better value from production subsidies and to foster an element of derived socio-economic benefit. In relation to 'near market' technologies whose cost structures are well understood, it is important to 'get the prices right' by using cost reflective support schemes. The flexibility and 'steerability' of policy instruments is therefore important. REFITs have allowed greater policy control because of a higher number of available instrument settings. High 'steerability' is attained by policy instruments which set cost reflective subsidies on the basis of technology differentiation and meaningful target-setting, leading to synergies in energy policy and industrial policy. But political will is also essential to redefine goals, alter the settings or indeed change instruments when necessary.

6
Integrating Wind Power into the Electricity Supply Industry

Introduction

Wind energy is at a cross-roads. Will it remain an atypical and marginal source, or is the sector heading towards 'normalisation'? For wind energy to become a mainstream source of electricity, its integration within the electricity supply industry (ESI) has to be improved. To clarify the ways in which national ESIs and the wind sector reciprocally influence the development paths of each other, this chapter explores its relationships within the ESI at three levels. The first section reviews the changing fuel mix in the electricity sector in order to establish the scope for displacement of conventional energy sources by wind. It analyses national ESIs, the market position of the utilities and the main consequences of liberalisation for incumbents and new entrants. The conduct and strategies of established actors inevitably impact on the scope for integration of wind power. Thus the second section reviews the economic issues of the market integration of wind power and the technical issues of its grid integration. The third section revisits the question of substitution of fossil fuels by wind energy from a different perspective, by considering the sector's capacity to achieve reductions in atmospheric emissions. Through these analyses, the chapter develops the argument that the 'normalisation' of wind power is underway but incomplete. Further reforms in policy and practice can be expected to reduce its atypical characteristics.

The impacts of national ESI configurations

National ESIs have underdone several rounds of restructuring in recent decades, with the most recent caused by liberalisation and the creation of a 'single market' in electricity within Europe. Changing contexts have reconfigured the opportunities and constraints for renewable energy sources. These will considered from several viewpoints. First, the composition of the fuel mix in the electricity sector is analysed. Second, the problem of energy gaps is identified and the scope to close gaps by fuel switching to renewables is explored. Third, the main consequences of liberalisation for both established actors and new entrants are reviewed.

Energy sources for electricity generation

Data on the structure of generation by fuel source in the five sample countries is provided in Table 6.1. Four of them rely heavily on fossil fuels. Coal provides approximately half of electricity generation in Denmark and Germany, and a third in the UK and Spain. Nevertheless, coal usage has reduced dramatically in Denmark and the UK, where in the 1980s it provided 80–90 per cent of electricity. The 'dash to gas' in the 1990s affected Denmark, Spain and to a lesser extent Germany but it is the UK system which became predominantly gas-based, through 'indigenous' North Sea sourcing. In France, usage of fossil fuels is a residual component at a low 10 per cent. The French ESI is dominated by nuclear facilities, whereas Denmark has none. The ESIs of the other three countries are relatively diverse in their range of sources. However, Spain has a moratorium on nuclear power, Germany has planned a phase-out on political grounds whilst the UK position has been ambiguous, with the nuclear option officially remaining 'open' but treated with scepticism, mainly on economic grounds. Yet in Germany, Spain and the UK, nuclear produces between a fifth and a quarter of output. Recourse to renewables is growing, with substantial proportions sourced from hydro, biomass, landfill gas and wind.

Table 6.1 Gross electricity production by fuel in percentages in 2004

	Coal	Gas	Oil	Nuclear	Hydro	Biomass/ Wastes	Wind/ Solar	Total
	%	%	%	%	%	%	%	%
Denmark	46.3	24.2	4.1	0.0	0.1	8.9	16.4	100.0
France	4.8	3.1	1.3	78.3	11.4	0.9	0.2	100.0
Germany	50.2	10.2	0.7	27.6	4.5	2.6	4.2	100.0
Spain	28.5	19.9	8.6	22.7	12.3	2.4	5.6	100.0
UK	33.6	41.3	1.8	19.2	1.9	1.7	0.5	100.0

Source: Electricity Information © OECD/IEA, 2005, part 1, table numbers 3 and 4, pages 36–7.

Table 6.2   Gross electricity production by fuel in TWh in 2004

	Fossil Fuels TWh	Nuclear TWh	Hydro TWh	Biomass/ Wastes TWh	Wind/ Solar TWh	Total TWh
Denmark	30.08	0	0.03	3.57	6.65	40.33
France	52.82	448.24	64.9	5.17	1.1	572.24
Germany	370.3	167.34	27	16.02	25.46	606.12
Spain	160.07	63.61	34.44	6.63	15.66	280.4
UK	294.12	73.68	7.41	6.7	1.82	383.73

Source: *Electricity Information* © OECD/IEA, 2005, part 1, table number 3, page 36.

However, data on relative proportions are only part of the story. To provide a fuller picture, Table 6.2 gives information on quantities of electricity generated by categories of fuel source. Firstly, in terms of aggregate generation, the largest producers by some margin are Germany and France, with Danish output representing a fraction of the other four. Secondly, French hydro generates a larger quantity of electricity from renewables than any other European nation, although wind generates a substantial number of terawatt hours in Germany and Spain. Thirdly, France generates more electricity from fossil fuels than does Denmark. In 2004, coal-fired generation alone was some 50 per cent higher in France than in Denmark, at 27.9 TWh and 18.7 TWh respectively (IEA, 2005b: I.37). Fourthly, the scale of the nuclear sector is worth noting. In France, nuclear power alone exceeds total generation in the UK and Spain, and is equivalent to eleven times total Danish production. These data provide a baseline for assessing the scope for fuel displacement, since wind energy may only displace what is already there. Although this fact may seem self-evident, its consequences are not.

Energy gaps and fuel switching

One of the major challenges for economic development is to source energy in the right quantities and the most appropriate manners. To close energy gaps, competition between fuels is long-established. The 'hegemonic battle' (Elliott, 2003: 185) between renewable and traditional energy sources is a recent phase of this competition. The biggest opportunity for switching to wind power is offered by ESIs dominated by coal-fired generation. There are two main reasons for this. One is that wind power can be integrated more easily into coal-dominated than into nuclear-sourced systems. The second is that coal is the most polluting energy source, so substitution by an emissions-free source such as wind achieves the greatest environmental benefits. Historically, several rounds of fuel switching in the ESI have occurred since the oil shocks of the 1970s, with the fate of coal-fired generation being a major question for all countries sampled. The stakes are high because each round of switching prompts restructuring in the electricity sector, with significant losses and gains of market share for interested parties. Next we review early phases of fuel-switching across the five nations in order to contextualise the scope for substitution by wind power.

Denmark in the early 1970s sourced 93 per cent of electricity from oil but, due to major price hikes, converted rapidly to a predominantly coal-based system (Mez and Midttun, 1997: 313). Whilst this switch provided a quick response to the oil crisis, it aggravated the problem of atmospheric emissions. The nuclear option was debated but dropped due to societal opposition. This generation mix proved uniquely favourable to the introduction of wind power. Fuel switching was accompanied by energy saving measures and greater use of district heating. Substantial increases in combined heat and power (CHP) introduced competition, weakened the utilities but also posed the danger of over-supply. In the early 2000s, about 40 per cent of Danish electricity consumption came from CHP and RES-E subject to a purchase obligation and paid at premium prices. These alternative technologies put the utilities' capacity to coordinate the electricity system under stress. Reforms of interventionist pricing mechanisms to manage over-supply led to the 'boom and bust' cycle in wind power investment discussed in Chapter 4.

In Germany, base load comes from brown coal and nuclear – being the cheapest sources – whilst load-following plants are largely hard coal and gas (Alt, 2005). Germany's fuel sourcing structure reflects the priorities of energy security and industrial policy. Domestic coal production has been a means to attain both, but causes large-scale atmospheric emissions. It is also loss making, with a long history of support by state subsidy (Bartle, 2004). The 1980 *Jahrhundvertrag* obliged the utilities to buy large quantities of German coal at production cost, whilst cheap non-EC imports were restricted. Support for coal was responsible for high electricity prices in Germany to which the *Kohlepfennig*, a levy on electricity sales, contributed (Haugland, Bergesen and Roland, 1998). Under pressure from the European Commission, these measures were phased out over 1995–6 (Mez, 1997). The reliance on coal has nevertheless continued, though alleviated by increasing recourse to gas imported from Russia. Meanwhile nuclear power has faced strong opposition from Germany's influential green movement, with a politically inspired phase-out now in progress. This configuration has proved highly favourable to renewables such as wind.

The political dimension of fuel competition has also been important in the UK (Man, 1987: 113). Britain had a coal-dominated electricity sector in the 1970s. However, the policy to support domestic coal production was reversed under the Thatcher government, provoking the miners' strike of 1983–4. The defeat of the miners' union and the decline in the coal industry, the development of oil and gas extraction in the North Sea, together with the liberalisation of the electricity and gas sectors, were the major factors encouraging fuel-switching. The 'dash-to-gas' in the 1990s led to the construction of 21.7 GW of CCGT capacity by 2001 (Starapoli, 2003: 64). It was mostly at the expense of coal, whose share of generation fell from 73 per cent in 1990 to 32 per cent by 2000 (MacKerron, 2003: 42). The switch from coal to gas also reflected a major transition from state to market. Yet the survival of nuclear power was engineered through state support. In a context of rapid depletion of North Sea gas, the option of further nuclear build was revisited during the 2005–6 energy review. This configuration has raised uncertainties about the evolution of electricity sourcing, provoking both hopes and fears regarding the future of renewables.

Spain during the closed and inward-looking Franco period aimed to be self-sufficient in industrial terms. Generating capacity was developed around indigenous sources, mainly coal and hydro. Spain increased its coal production from 10 to 19 million tonnes between the 1970s and 1985 (with modest contraction thereafter), whilst net imports rose from 2.3 to 11 million tonnes between 1970 and 1994 with 80 per cent of coal use being in the electricity sector (Haugland, Bergesen and Roland, 1998: 85). Restructuring occurred in the 1990s – with closure of uneconomic mines and employment losses – but major subsidies to the coal industry were phased out only in 1998 (Serrallés, 2005). Like the UK, Spain has seen a major increase in CCGT capacity and has a significant nuclear fleet, currently comprising eight plants. Unlike the UK, Spain has no oil or gas reserves and coal supply is limited – yet energy demand has been growing rapidly. Thus the core problem faced by Spain's energy-hungry economy is high dependence on imports – at the 80 per cent level in 2004 – raising issues of energy insecurity and economic vulnerability, especially in periods of increasing oil and gas prices. This context has been favourable to development of renewables as an indigenous resource.

France, like Spain, has very limited conventional energy reserves and so been preoccupied by 'energy independence'. The drive to cut petroleum imports after the oil price shocks was the main factor behind the huge expansion in nuclear-sourced electricity in the 1970s and 1980s, with the building of 58 nuclear reactors. However, supply outstripped demand, leaving France with structural over-capacity. This problem has been managed by exporting around 15 per cent of generation annually. Meanwhile fossil fuel generation was mostly phased out in mainland France, with remaining plants used largely to meet winter peaks (though the overseas territories have specific characteristics). France is also the European nation with the largest quantity of renewables generation, but almost exclusively from hydro. This unique configuration of dominance by nuclear power, large-scale hydro generation and structural over-capacity has been unfavourable to 'new' renewables such as wind.

The development paths of national ESIs help explain the nature and intensity of renewables policy in the recent period, with the influence of industry contexts shading from the favourable in Germany and Spain over to unfavourable in the UK and France, with Denmark in the middle (being a pioneer now characterised by a mixed investment environment). But further structural developments within national ESIs over the next decade can open major opportunities for RES-E. One possibility is that coal follows oil in being phased out of mainstream electricity generation. But whilst coal has seen massive reductions and the trend to downscaling may persist, its disappearance – even in Europe – is unlikely since it is a cheap and plentiful energy source. A crucial question is whether the current generation of advanced coal-fired technologies will provide sufficient efficiency gains and environmental improvements to secure the industry's near-term prospects. An unanswered question for the long term is whether a technological revolution will allow emissions from coal to be managed economically and reliably by carbon capture and storage. Question marks also hang over the future of the nuclear industry. Will Germany's phase-out be confirmed or reversed? Will the UK opt out of or back into nuclear power? Whilst some uncomfortable uncertainties persist, these contexts provide openings for renewables. Thus Gipe (1995: 479) called for 'sustained orderly development' to replace aging, inefficient and polluting power plants with RES-E installations.

Where then are the sourcing gaps? In Germany, a key component of ESI restructuring is the 2002 law programming the decommissioning of 23.5 GW of nuclear capacity.[1] This politically motivated phase-out was agreed under pressure from the antinuclear Green party, which was the major coalition partner of the SPD in the Schröder administration. Moreover, some 17 GW of coal-fired capacity will also need replacement or upgrading. In principle, as much as 40 GW from a total capacity of 121 GW could be decommissioned by 2020. Major new build will to required to fill this gap. Similar developments are predicted for the UK. According to the House of Commons Environmental Audit Committee (2006: 3): 'by 2016, it is likely that between 15 and 20 GW of electricity generating plant will be decommissioned. This amounts to nearly a quarter of UK generating capacity'. Although uncertainties persist regarding the impact of the new EU Large Combustion Plant Directive, perhaps 10 GW of coal plant will disappear. In the nuclear sector, a reduction of some 11 GW of capacity is expected by 2023 (House of Commons Environmental Audit Committee, 2006: 10). Also, demand is growing by around 1.5 per cent per year. In France too the nuclear fleet is showing its age. In Spain, question marks hang over the renewal of its nuclear sector, given the existing moratorium. In all these countries, a substantial 'generation gap' is set to progressively open over the next twenty years. On the other hand, the challenges facing Denmark are rather different. It has no nuclear facilities to worry about, but its dependence on polluting coal remains a problem, especially as a substantial proportion of large-scale fossil fuel generating plant and CHP is of recent vintage.

In some countries, decisions on new build are in the pipeline, but much remains to play for. In Germany, new conventional plants are already being built, including coal-fired generation. As noted by the (IEA, 2002: 8): 'the German government wishes to maintain a significant coal-based electricity generation capacity to avoid over-dependence (...) on imported energies'. But whether recourse to coal and gas are full and viable means to close sourcing gaps is questionable. In France, the 2005 Energy Bill reinstated the preference for nuclear power. The lifespan of existing reactors has been prolonged and a demonstrator European Pressurised Reactor plant is scheduled for construction in France by 2012, with a view to rebuild the nuclear fleet after 2015. In the UK, an 'energy review' was conducted by the DTI in 2005–6. A range of options are being considered, including the construction of new nuclear reactors, but with no firm decisions to date. Spain and Denmark have yet to sketch long-term evolution paths for their ESIs, apparently defaulting to incremental adjustment.

Thus sourcing gaps create opportunities for renewables across a range of scenarios. The aspirational target of 20 per cent electricity generation in renewables by 2020 expressed by Germany and the UK, together with even higher targets by France, Denmark and Spain, may represent a major shift of resources between energy sources and conversion technologies. Although increases in energy demand make it possible that the growth in RES-E market share does not translate into an absolute loss of sales for conventional sources, the 'energy race' will probably produce winners and losers. Renewables producers are jockeying to be the winners by exploiting closures in conventional generation capacity. The coal and nuclear sectors are fighting back to avoid being the losers.[2]

Liberalisation and market development

The major cause of ESI restructuring in the recent period has been the liberalisation process which has changed the shape of markets, the strategies of incumbents and the opportunity structure for new entrants. Ironically, national policies to support renewables and new suppliers have required extensive interventionism. This has resulted in continuing tensions between state and market, and uneven outcomes from liberalisation.

The UK under the Thatcher government was a pioneer of electricity liberalisation. The change was highly politicised and concentrated on the ownership dimension. Although market competition in electricity was possible with publicly or municipally owned companies as shown by the Norwegian case,[3] the British Conservative position was characterised by deep distrust of public ownership and state-centred planning. Privatisation became almost an aim in itself. The transition began with the 1989 Electricity Act. By 1999, full liberalisation had been achieved, with separation of grid ownership from generation and trade. The liberalisation process resulted in several waves of merger and takeover. Whereas 12 regional electricity companies were set up in 1989, by 1997 ten had been taken over, seven by US companies (Young, 2001: 99). The duopoly in generation exercised by National Power and PowerGen has been replaced by the market domination of six multinational suppliers providing over 70 per cent of electricity. The major players in England and Wales in 2004 were RWE AG (formerly National Power/Innogy) with 21 per cent of the market, British Energy (18 per cent) and PowerGen (17 per cent) (Serrallés, 2005). The transmission network is owned and operated by National Grid Transco, with responsibility for the whole of the national grid. In Scotland, two vertically integrated utilities – ScottishPower and Scottish and Southern Energy Supply – hold 98 per cent of the generation market. Separate arrangements apply for Northern Ireland.

In contrast to the UK, the Danish ESI was traditionally highly decentralised, with a large number of municipal companies and a strong public service ethos. Some 65 per cent of the system was owned by consumers; companies were not allowed to make a surplus, but neither did they pay taxes on profits or assets (Midttun, 1997: 295). The 1999 Danish Electricity Supply Act restructured the ESI by an unbundling of generation and grid ownership. By 2005, major consolidation in the supply industry was underway with a merger between the main Danish electricity utilities, Elsam and Energi E2, involving a partial divestment to Swedish Vattenfall (IEA, 2006b: 31).

In Spain, liberalisation of the electricity sector began with legislation in 1994 initiating the unbundling of the transmission grid and setting up an independent operator, Red Eléctrica de España (REE). The 1997 law accelerated the process, allowing consumers to choose their electricity supplier. Liberalisation was the catalyst for a wave of mergers, with 30 independent regional generation firms disappearing between 1990 and 2002, leaving five main suppliers. Endesa, Iberdrola and Unión Fenosa alone accounted for 83 per cent of generation in 2004 (Serrallés, 2005). Their strong home base has allowed them to diversify their product portfolio (into gas and mining, construction and telecommunications) and to expand into global markets.

The main utility in France is EdF which in the 1990s covered 94 per cent of output and 95 cent of the transmission and distribution grids (Poppe and Cauret, 1997: 205). Although previously publicly owned, it has been partially privatised. EdF is the dominant giant in French and European electricity markets, but it never had a monopoly of domestic supply. The Compagnie nationale du Rhône operates hydroelectric plants, whilst other generators included the Compagnie générale des eaux and Société lyonnaise des eaux. France has a number of international connecters, allowing EdF to export electricity to Belgium, Germany, Italy, Spain, Switzerland and the UK. In effect, EdF stabilises nuclear base-load generation by exporting its structural surplus, whilst importing smaller quantities of load-following electricity where necessary.

In contrast to France, the German ESI was primarily privately owned and enjoyed a high level of autonomy from the state (Sturm, 1996: 129). The Energy Industry Act of 1935 resulted in a decentralised but cartelised system, involving monopolies within localised territories in generation, transmission and distribution (Wüstenhagen and Bilharz, 2005). Because of revenue-sharing through concession fees paid to municipalities by utilities, Mez and Midttun (1997: 315) argued that the German situation constituted a 'case of regulatory capture and state failure'. As a result of their privileges and structural position in the economy, Mez (1995: 180) concluded that 'public utilities belong to the most powerful interest group in West Germany', with considerable influence on policy formulation. By the 1990s, the West German system was based on eight distinct but federated systems, characterised by different levels of vertical integration (Mez, Midttun and Thomas, 1997: 5). In 1998 the German electricity and gas sectors were liberalised. The new law required electricity companies to unbundle their activities and accounts, gave legal rights for third party access and freed consumers to change suppliers. This was expected to break up the old monopolies. Although liberalisation led to large-scale restructuring, the utilities have nevertheless consolidated their position by takeovers and mergers. The number of major generators was reduced from eight to four – RWE, E.ON, EnBW and Vattenfall Europe. These firms have also split the German regional grids into four monopolies, of which they each run one. In the early 2000s, the historical operators reinforced their dominant position through the erection of barriers to entry by a reduction in generation prices accompanied by an increase in grid access costs, in a context where no regulator existed as yet (Larcher and Revol, 2003: 40; Wüstenhagen and Bilharz, 2005).

In summary, liberalisation has broken up publicly-owned monopolies and separated generation from transmission and distribution activities. Yet takeovers have led to industrial concentration and constitution of international giants, several of whom dominate more than one national market. This process has created opportunities for new energy sources and providers, but has also introduced new constraints. In a context of incomplete liberalisation, conflicts have arisen between historical operators wishing to preserve market share and new energy purveyors seeking to transform existing markets. The creation of a European electricity market has been impeded by the limited physical interconnections between national markets (discussed further below), and complexities in the setting up of a European trading system. The EWEA (2004) argued that market failures lead to discrimination against renewables. The main distortions are subsidies to conventional sources, non-internalisation of external costs and grid infrastructures tailored to conventional, centralised generation plants. The consequence is that innovations in supply can be held up by the inherent characteristics of renewable energy sources, by the structural constraints of grid design and by the conduct of historical operators and their lobbies. These factors pose major integration challenges for wind power, which will next be reviewed.

The economic and technical challenges of integrating wind power

The integration of wind power into national ESIs has posed its own technical and economic problems. Conventionally-sourced electricity is 'dispatchable', meaning that programmed quantities can be delivered by increasing or decreasing generation at will. However, wind energy is one of several renewable sources based on natural 'flows' rather than artificial 'stocks'. Its output can be turned down (by curtailment), but not up. Wind speeds are in constant change, resulting in stochastic fluctuations in electricity output usually termed as 'intermittence'. On the other hand, conventional generation sources do not display infinite flexibility either. Ramp rates (the ability to increase or decrease generation) vary significantly, being high for hydro and gas but low for nuclear. All electrical equipment requires scheduled maintenance and can suffer unscheduled failure. Utilities and grid operators have, however, considerable experience in the anticipation and management of constraints posed by conventional sources.

System operators match the production and consumption of electricity in real time, since storage remains a marginal option. Grid operators use sophisticated techniques to anticipate demand, and match it by drawing on different categories of 'base load', 'load-following power' and 'balancing power'. The 'base load' is the load that is permanently required day and night throughout the year. Overall consumption is higher in winter than summer, and greater in the day than the night. This creates seasonally distinct diurnal cycles marked by predictable peaks during which all users require electricity, followed by lower-usage periods where some users decrease or stop consumption. 'Load-following power' tracks predictable developments, whilst 'balancing power' can be considered as 'fall back' reserves of generation to meet unplanned contingencies, such that demand and supply are always in balance and correct electricity frequencies are maintained.[4] The three categories of power are made available, grid stability assured and demand satisfied by production schedules arising from bid prices made by generators to the market and transmitted to the grid operator. This operational strategy has a history of responding successfully to rapid variations in demand, frequency changes and breakdowns, with blackouts being rare in Western Europe.

Wind power raises new and additional challenges. In the absence of substantial storage, wind energy cannot provide a stand-alone electricity source. Whilst it is possible to construct a fully-functioning electricity system with coal or gas plants alone, it is impossible to rely just on wind turbines. This has consequences both for the ESI and the wind sector. Output of electricity from wind installations is usually too high or too low to match local demand – where local may be a house, a factory or a community. The solution is to sell surplus production into the grid and draw power from it when wind is not available. But as flagged by Jørgensen and Karnøe (1995: 76) 'the idea of local energy production satisfying local needs has been undercut not only by economic interests but also by the unstable production from wind turbines that necessitates external power supply or connection to a network'. Thus wind power has a *symbiotic relationship* with conventional generation. This puts into question a conceptualisation found in the discourse of enthusiasts that wind power overcomes dependence on conventional sources. In practice, dependence is reduced but not eliminated. On the other hand, the utilities have sometimes stressed a rival conceptualisation of wind power as a mere 'fuel saver'. In this vein, E.ON Netz – the TSO with the greatest amount of wind power in its grid in Germany – commented that 'due to their limited availability, wind power plants cannot replace the usual power station capacities to a significant degree, but can basically only save on fuel' (E.ON Netz, 2004: 7). Further, wind power's reliance on conventional back-up can allow utilities to preserve their dominance.

The balancing power required to compensate for intermittence in wind availability is an adjustment variable related to supply, unlike balancing power for load-following which is an adjustment variable related to demand. The former thus poses a new and distinct challenge for grid management, for whose handling grid operators had little experience in the past. Moreover, the capacity to make accurate forecasts is higher for diurnal demand cycles than for wind availability.[5] The shortfalls and surpluses arising from wind intermittence impact on delivery schedules and entail costs, raising the question of who pays for them. In northern Europe, these costs were mainly passed on to incumbents. These factors explain the historical reluctance shown by utilities to incorporate wind power.

The scope for displacement of conventional fuels by wind depends, to an appreciable extent, on the configuration of national ESIs. In the view of DeCarolis and Keith (2005: 73) 'all else equal, the cost of intermittency will be less if the generation mix is dominated by gas turbines (low capital costs and fast ramp rates) or hydro (fast ramp rates) than if the mix is dominated by nuclear or coal (high capital costs and slow ramp rates)'. In practice, the potential for substitution is therefore constrained. As stressed by the IEA (2002: 122) 'generation from renewables tends to be intermittent and not suitable for base load'. This points to a mismatch between green hopes and technological reality. Radical greens would like to see the nuclear sector closed down, and offer wind as an alternative. But this is unrealistic, as nuclear power-stations provide base load for which wind power is unsuited. On the other hand, the case for wind power as a substitute for coal-fired generation is strong, *provided* that enough is present to displace readily and that there is adequate back-up capacity offering high ramp rates. At the same time, any displacement reduces sales by conventional generators and impacts on their economic viability. This results not only in competitive tensions but also conflictual relations between independent wind generators and incumbents. A relationship of complementarity between energy sources and conflict between types of electricity providers is a core reason why the deployment of wind power has necessitated policy intervention and the establishment of 'political markets'. The acceptability of wind power in the eyes of historical operators was consistently low. It is reasonable to assume that they would never have implemented the technology on their own initiative, but accepted it only as a marginal contributor through imposition (Denmark, Germany) or buy-off (Spain, UK). The question for the future is whether wind power will be accepted as a mainstream energy source on economic and technical terms comparable to the rest of the sector.

Market integration

In a fully liberalised electricity market, public authorities set the 'rules of the game' and supervise their implementation. Otherwise they leave market actors to 'play out the game', without further intervention. Market integration involves electricity producers generating revenue by making successful bids to the market, whilst accepting an equitable apportionment of costs related to grid access and balancing services. Thus the question of market integration explores the relationships (a) between categories of electricity generator in competition with each other, and (b) between generators and the system operator (who is responsible for the moment-by-moment operation of the electricity system).

But in contrast to the turn to liberalisation, the promotion of wind power through feed-in tariffs in Denmark, Germany, Spain and France has involved direct intervention in the electricity market. The most widely recognised category of intervention relates to the fixing of tariffs, which sets a 'floor' price for wind and removes income risk. In many cases, the 'floor' price has also been a 'ceiling' price because generators do not make bids to the spot market at times when supply tensions would allow them to fetch higher prices. A second, less widely recognised, category of intervention relates to the supply contract. As practised in Germany and latterly in France, the system operator is under a purchase obligation to accept all specified RES-E production. This reduces risks for generators, by obviating the need for a negotiable power purchase agreement. A third category of intervention is in relation to payments for balancing services. In the German, Danish and French systems, these are not paid by the RES-E generator, but by the TSO. Several consequences arise from these arrangements. The generator keeps all the feed-in tariff. This consolidates revenue and makes prediction of the income stream easier, both of which are advantageous for access to investment loans. Meanwhile the TSO picks up the services bill and passes it on to consumers.[6] This hidden payment increases the overall cost of renewables to end-users, over and above the feed-in tariff *per se*. In addition, responsibility for the provision of balancing services goes to the TSO. But this also means that wind operators have no responsibility for managing their level of output. Fluctuations in wind-sourced generation and resulting mismatches between supply and demand become a problem solely for the TSO. From the point of view of the utilities, these privileges accorded to wind power distort competition by taking market share from other generators and by altering the responsibilities of the system operator. Because of the financial implications, tensions between incumbents and the wind sector were inevitable. Preservation of the feed-in system therefore required serial renewal of state intervention. As seen in Chapter 4, political will to do so was maintained in Germany but faltered in Denmark in the early 2000s. In Germany, a highly interventionist support system has continued, with the result that the market integration of the wind sector is low. In Denmark, policy reforms have sketched out routes to greater market integration. Whilst older wind installations benefit from a diminished version of the previous system, new entrants make offers to the spot market. This is a substantial hurdle for small entrepreneurs, and contributes to the current stalemate.

By contrast, the UK was not only the earliest European country to liberalise the electricity sector, but also the one which sought most consistently to construct a 'pure' electricity market. The New Electricity Trading Arrangements (NETA) operating in the early 2000s were designed to penalise unpredictability and created obstacles for generation from renewables, especially wind (OECD, 2002: 143; IEA, 2002: 53). The RO gave wind power none of the privileges accorded under the German feed-in tariff. Wind operators must accept the risks related to variable prices, negotiation of power purchase contracts and the cession of part of their ROC income to pay for balancing services. In the process, the UK system achieved a high level of market integration for renewables, but at the cost of delegating their deployment to a single category of entrepreneur – large firms, including the utilities – and made them the major beneficiary of consumer subsidy. The 'private' sector that is so important in Germany and Denmark has been squeezed out because, with few exceptions, only major players can cope with the complex, risk-laden UK environment. Intra-ESI conflicts are certainly reduced, but only through the erection of barriers to entry to small-scale investors.

The Spanish support system in its 2004–6 version provided an interesting 'halfway house' between the state intervention inherent in REFITs and the market integration sought by RPS systems. In the 'market option' (discussed in Chapter 4), the generator made offers to the spot market and received the pool price. In addition, subsidies were paid for market participation, whilst wind farms above 10 MW must forecast production for the day ahead. Deviations from forecast greater than 20 per cent incurred penalties. Operators were therefore incentivised to take responsibility for output levels. To improve forecasting accuracy, a major study was orchestrated by the AEE in 2004–5 which concluded that improvement was possible but, even with state-of-the-art methodologies, the lower limit on errors would still be around 25 per cent (AEE, 2006: 81–7). Recommendations for improvement included metrological predictions on a smaller scale and the 'pooling' of wind farms to cancel out errors. The pool approach is facilitated by the large size of Spanish wind farms and an ownership structure in which utilities are predominant. Grid management improved as a result, with the requirement for balancing services dropping from 1336 GWh in September 2004 to 767 GWh in September 2005, despite increases in wind output (AEE, 2006: 71). An associated innovation was the proposed establishment of local control centres for wind farms above 10 MW, organised on the basis of company ownership or geographical area (AEE, 2006: 33). Once operational, these will be coordinated by the grid operator *via* a national control centre, known as CECRE. The long-term aim is to manage wind power in a manner closer to conventional dispatching, and so reduce intermittency problems. National control, however, decreases the local, 'embedded generation' aspect of wind power and increases centralisation.

Spain has sought to combine the effectiveness of feed-in tariffs with promotion of greater market integration, with the result that by 2006 all but the smallest producers had chosen the 'market option'. However, the trend to utility domination of the wind sector have been reinforced by market-based support in both Spain and the UK. In Denmark and Germany, large-scale wind deployment has brought the lack of market integration – and its problems of unpredicted and uncontrolled generation – into sharper focus. Further, if the aspirational targets of 20–30 per cent of electricity from renewables are met, it is improbable that such a large slice of generation can lie permanently outside the market. The 'Spanish solution' offers a path for exploration. Yet it cannot be transplanted wholesale to Germany or Denmark, as the ability of small-scale producers to combine in larger consortia is inherently lower than for Spanish utilities, whilst their willingness to accept incentives and penalties for delivery to forecast is uncertain. Market designs, however, can be tailored for national configurations. The key to future wind power integration will lie in tailoring national choices on market construction to evolutionary developments. One possibility would be to offer phased incentives for a pooling of wind farms to improve forecasting and coordinate dispatching to ease the way towards a more market oriented system.

Grid integration

Improved management of wind power flows raises the question of grid integration. According to Chris Shears, the BWEA chairman, getting wind power to users 'is the biggest strategic issue that we face and the most complex' (Massy, 2005: 52). As the penetration level of wind power has risen, its integration into the grid has catalysed problems of several orders.

Grid connection issues

The question of how to obtain grid connections and who pays for them has arisen in all the reference countries. Areas characterised by high wind speeds are often located away from load centres, at points where grids are 'weak' (IEA, 2002: 54). Apart from limiting the availability of connections, this factor also funnels development into stronger sections of the grid, reinforcing a tendency to regional concentration (discussed in Chapter 7). The consequences are increased competition between developers and long grid queues, slowing the rate of wind power deployment, together with a tendency to 'cumulative effect', provoking local opposition.

In France, which as a late-comer to wind power might be expected to have few problems, obtaining a grid connection can take several years

(Boston Consulting Group, 2004: 24). By 2003, there were some 14,000 MW of mooted wind farm proposals, whereas only 6000 MW of connections were considered feasible before grid reinforcement was undertaken (Chabot, 2003: 17). In the UK, a shortage of connections has caused a backlog, leading Massy (2005: 52) to claim that 'much of the 10 GW of wind projects hoping to connect to the system will have to wait for grid upgrades'. In the wind pioneer countries, a lack of connections is contributing to saturation problems. With regard to Germany, Luther, Radtke and Winter (2005: 236) noted that 'in the northern part of the power system, the transmission capacity of the high-voltage system reaches its limits during times with low load and high wind power production'. In Denmark, onshore expansion is stymied. In Spain, wind capacity targets set by the regional authorities are double those set by national authorities (AEE, 2006: 53), and cannot be met due to lack of grid infrastructure. A shortage of connections has often prevented small projects going ahead (Dinica, 2002: 221–2), reinforcing domination in Spain by large-scale utility projects.

Grid management of wind power flows

In the pioneer countries, a further problem relates to the management of wind power flows, nationally and internationally. With some 7000 MW of wind in its control area, grid instability remains a worry for E.ON Netz, the north German TSO, who stated 'there is therefore a risk that even simple grid problems will lead to the sudden failure of over 3000 MW of wind power feed-in (...) At the present time, it is not known how to confront this risk' (E.ON Netz, 2005: 22). The German Energy Agency acknowledged that 'in certain situations (strong wind and low load), Germany sees a surplus in power generation on a few days per year. In such situations, huge power flows to neighbouring countries can be observed (...) Large-scale power generation from wind energy converters in Germany impairs considerably the reliable operation of the grids in neighbouring countries' (DENA, 2005: 10). The European Commission (2005: 44) observed that:

> the influence of wind power on cross-border bottlenecks between Germany and its neighbours has created some disturbances in the Netherlands and Poland. Arrangements for power plant scheduling, the possible rigidities of the electricity market, reserve capacity for cross-border transmission and congestion management seem to be crucial points requiring further analysis.

Improvements in Germany's grid management of wind power are therefore essential.

In Denmark, the concentration of wind installations in Jutland has highlighted the country's problematic grid configuration. Denmark has two independent grids – one for the west and one for the east – with no interconnections between them. The consequence is that wind power generated in Jutland which is surplus to local needs cannot be transmitted to Copenhagen or other eastern load centres. With an aggregate 2400 MW of international connectors, surplus wind power from west Denmark is exported to Norway, Sweden or Germany to help balance the western grid (Sharman, 2005). On the other hand, Spain cannot adopt this solution because it has limited interconnectors with neighbours. Being essentially an island system, grid stability in Spain has to be managed domestically. REE, the grid operator, has repeatedly warned of the dangers of large wind power flows and aimed to place a ceiling on capacity. To mitigate the problem, Spain has taken a lead in drawing up rigorous grid access codes to govern the technical characteristics and behaviour of turbines, with particular attention to so-called 'voltage dip ride-through capability'. The aim is to ensure that wind turbines do not trip off-line during a voltage dip, triggering further grid instability and black-outs. As an island system, the UK is faced with similar challenges. Britain must learn to manage its wind power domestically, with the added complication that the interconnector between England (where the main load centres lie) and Scotland (where most new wind deployment is taking place) is currently small.

Grid reinforcement and reconfiguration

Solutions to the infrastructure problem have two main dimensions. One is to reinforce sections of the grid. The other is to undertake grid reconfiguration. In both cases, the distribution of costs between grid operators and developers / generators creates tensions, with each wishing to pass the burden to the other. Grid reconfiguration is, by definition, a more ambitious undertaking. Existing grids reflect an historical relationship between the development of load centres (cities and industrialised zones) and the location of power plants. Future grids may require development in rural and offshore areas, with transmission over long distances to existing load centres.

In Spain grid bottlenecks have restricted wind power expansion (Río and Gual, 2006). The ambitious targets in the 2005 PER entail grid reinforcement, notably in western Andalusia, Aragón, Catalonia and Valencia (AEE, 2006: 102–3). Larger interconnectors with France,

Portugal and Morocco are also envisaged. In Denmark, interconnection of western and eastern grids is planned for 2010 (Børre Eriksen; Akhmatov and Orths, 2006: 216). In Germany, wind generation is concentrated in the northern and eastern *Länder*, whilst the main load centres are in the west (Essen-Cologne-Bonn) and south (Munich). As capacity and output increase, the principle of using wind power close to the point of production becomes less feasible, necessitating national transmission. To remove bottle-necks, some 850 km of new HT lines are scheduled for completion by 2015 (DENA, 2005: 9). For the long term, the challenges are even greater, given ambitious plans for 20,000–25,000 MW of offshore wind capacity by 2030, all in the north.[7] The transmission of the proposed 30 TWh of offshore electricity to inshore conurbations could entail grid reconfiguration, major expenditure and potential for power losses. The impacts of HT lines on landscape, nature, and housing are expected to lead to planning objections and delays. In Scotland, grid reinforcement is proposed for the Beauly-Denny line. The feasibility of laying undersea transmission cables connecting Scotland and north-west England has been investigated (IEA, 2002: 54; National Audit Office, 2005: 76). For Scotland, the issues raised by large-scale wind deployment – onshore, offshore and on the islands – probably combined with marine generation technologies in the next decade are comparable in scale to those faced by Germany. Grid reconfiguration illustrates the consequences of mixing 'hard' and 'soft' paths. New patterns of industrialisation are involved, with a radical reworking of logistical and financial parameters. In the process, the concepts of 'centralisation' and 'decentralisation' may be stretched, remodelled or simply lose their meaning.

Wind power, emission reductions and national ESIs

Renewables are often proposed as a response to the problems of climate change (Waller-Hunter, 2004; Holttinen and Tuhkanen, 2004; Sjödin and Grönkvist, 2004). The European Environment Agency has even stated that 'the largest emission savings for EU-15 are projected to come from renewable energy policies, followed by the landfill directive' (EEA, 2004: 22). Around Europe a substantial proportion of GHG emissions come from power stations; in the UK, they account for about a third of CO_2 emissions. This context has encouraged wind lobbyists to make bold claims regarding the capacity of wind power to effect emission reductions. The EWEA and Greenpeace (2002: 12) stated that 'a reduction in the levels of carbon dioxide being emitted into the world's atmosphere is the most important environmental benefit from wind power generation'. Arthouros Zervos, now president of the GWEC, asserted that 'wind energy is today one of the cheapest options in reducing CO_2 emissions' (Zervos, 2003: 319). However, the justification for such claims merits close investigation if we are to understand the causal pathways to emission control and to act on that knowledge effectively and efficiently. The starting point for this investigation is the observation, based on circumstantial evidence and practitioner reports, that wind power already makes an important contribution. But there is still a need to clarify the causal pathways by which this occurs and quantify the extent of its occurrence accurately.

Wind power and the mechanisms for GHG cuts

It is sometimes believed that it is the recourse to wind power – or other GHG-free energy sources – which cuts emissions. This is a misunderstanding. The generation of electricity from wind energy has the great merit of producing zero emissions in operation – no GHG, no SO_2 and no NO_x. But because it has no emissions, the application of wind technology can only cut emissions *indirectly*, namely by displacement of fossil fuels. What happens elsewhere in the ESI is therefore as important as what happens with wind power. Consequently the issue of emission control is inextricably linked to the discussions above on fuel substitution and the symbiotic relationship between wind power and other generators.

The high road to cutting GHG emissions from fossil fuels is simply to burn less, in other words to cap and then reduce consumption. Other pathways are to substitute lower for higher emission sources (e.g. gas for coal, and so forth), and to practice sequestration and storage at the point of emission (once the technology is proved viable). Thus a reduction in fossil fuel combustion can arise from (a) a lower level of energy demand, (b) greater efficiency in use and (c) substitution by non-fossil fuels. For wind power to *reduce* atmospheric emissions by substitution, the following conditions must all be met:

1) electricity from wind energy must directly displace generation from a fossil-fuel source (rather than substitute for a non-GHG-emitting source);
2) the recourse to balancing power due to intermittence should not result in a net increase in fossil-fuel emissions;
3) an increase in electricity demand, where it occurs, should be lower than the increase in electricity supply from wind (and other non-GHG sources).

In other words, if wind power displaces an electricity source which has minimal or zero emissions, there is little or no emissions cut. This can occur in the Nordic pool when surplus wind power from Denmark is exported and substitutes for Norwegian hydro. Secondly, in ESIs with a large fossil fuel sector the recourse to balancing power is highly unlikely to contribute to additional emissions, but the situation is less clear where – as in France – the ESI is largely GHG-free already but based on inflexible nuclear power. Thirdly, a large increase in demand for electricity can lead to a situation where supply from *both* fossil-fuel and non-GHG emitting sources increases. This cancels out the emissions reduction. It has happened in Spain, where expansion in wind power has been outstripped by even greater increases in electricity demand and associated rise in fossil fuel combustion. Finally, an additional complication in calculating emission reductions is that aggregate data on trends in a national ESI are not enough to identify the *causes* of decreases (or increases). Increased use of renewables will contribute to a downwards emissions trend, but so too will displacement of coal by gas or nuclear, as will greater efficiency, or reduced demand (say from favourable weather). Empirical identification of the circumstances under which wind power cuts emissions and in what quantities is therefore a complex matter.

The accountancy of emission reductions

Given the high-profile media coverage of climate change, it might be imagined that sophisticated accountancy methods exist to calculate the effects of GHG-free generators displacing GHG-emitters within national ESIs. In fact, no commonly agreed system exists, either in terms of principles or empirical methodologies. The House of Commons Environmental Audit Committee (2006: 49) commented that 'it is an extraordinary fact that nowhere in readily available Government documents could we find a published breakdown of emissions from the electricity generating sector by fuel type'. The full range of reasons for this inadequacy can only be conjectured,[8] but one reason can be advanced with confidence. This relates to the *time-bound* effects of weather-dependent generation sources, of which wind power is one example. Nowhere have wind farms led to the closure of conventional plants, since they only generate electricity when the wind blows. Whereas it is relatively straightforward to calculate changes in emissions levels when a coal-fired plant is closed and replaced by gas or nuclear, it is problematic to quantify what happens within the electricity system when wind power plants are grafted onto existing supply, given the symbiotic relationship identified above. In consequence, there is no validated system for the calculation of indirect emissions reductions caused by wind generation.

The emissions accounting system favoured by the wind lobby is based on the assumption that the displacement effect is constant in relation to a particular fuel. For wind power, the most advantageous source to displace is coal, since coal-fired generation has the greatest emissions per kilowatt hour. The wind lobby's accounting method assumes that every kilowatt hour of wind power displaces a kilowatt hour of coal-fired generation. This method is often used in the UK, where an emissions factor of 860g of CO_2 equivalent per kWh is typically assigned to coal. Annual emissions 'avoidance' by wind is calculated by multiplying anticipated generation levels by the assigned emissions factor (see Box 6.1 for commentary on the concept of 'avoidance'). The cherry on the cake is to multiply the result by 25 years, which is the conventionally-assigned lifetime of a wind turbine, in relation to the total capacity of a wind farm – or indeed the whole of the wind sector.

But what is the status of these calculations? An important characteristic is that they offer *predictions*. They forecast aggregate outcomes over the long term, based on a number of assumptions. The most important of these is that every kilowatt of wind power replaces a kilowatt of coal-fired generation – for 25 years. For this to happen, the primary condition is that sufficient coal-fired generation exists to allow displacement. This condition is mostly – but not always – fulfilled in the national electricity systems surveyed. It is certainly the case in Denmark and Germany, whose ESIs are still dominated by coal, whereas coal is less prominent in the other three.

But how does diversity of sourcing affect the scope for displacement of coal by wind? Expert opinion is divided on this question. In relation to the Nordic pool, Sjödin and Grönkvist (2004: 1559) drew up a 'merit order' of generating sources based on least variable-cost dispatch. In the cases of Denmark, Finland and Germany, coal was the most expensive source. This suggested that coal was most likely to be displaced by wind, with a typical emissions factor of 930g CO_2/kWh (due to the grades of coal used). Likewise Holttinen and Tuhkanen (2004: 1644) considered that coal had the highest marginal costs and would, in near-term scenarios for Denmark and Finland, be displaced by wind with an abatement level of 800–900g CO_2/kWh. However, they also noted that 'there might not exist old coal plant capacity for the whole wind power production to be replaced at all times of the year. During

Box 6.1   On the differences between emission 'reductions' and 'avoidance'

In English language discussions, five terms are commonly used to report downward emissions variation namely: 'reductions', 'cuts', 'avoided emissions', 'emissions savings' and 'offsets'. Regrettably, these terms are sometimes conflated and used interchangeably. This is a serious error, because the meanings are not identical.

The meaning of GHG 'reductions' and 'cuts' is that in relation to a baseline calculated for a specified date, the total quantity of emissions falls at some later point in time. This approach is exemplified by the 1997 Kyoto protocol, which aims for an aggregate GHG reduction of 5.2 per cent in relation to the 1990 baseline to be achieved over the 2008–12 commitment period by signatory countries. To become operational, the 5.2 per cent target has to be disaggregated to national, sectoral and subsectoral levels – for example by considering emissions for the UK, then within the energy sector as whole, and then within the subsector of electricity generation. Having specified the unit of analysis, a quantitative baseline must be established for that unit. A 'reduction' occurs when there is a sustained fall in relation to the baseline. The semantic constituents of the terms 'reductions' and 'cuts' can therefore be specified as (a) a quantified and verifiable fall in emissions (b) which has occurred in relation to an explicit baseline (c) during the recent past.

On the other hand, these semantic constituents do not enter into the meaning of the terms 'avoidance', 'savings' or 'offsets'. Indeed, usage of these terms often displays creative ambiguities. In relation to wind power (or other comparable technologies), the primary meaning of 'avoidance' is that no emissions are produced. But a frequent unspoken assumption is that, in the eventuality that wind power had not been deployed in a particular instance, a GHG-emitting source would have been used to generate the same quantity of electricity. The problem is that the 'avoidance' term transports us to a world of 'counterfactuals' or 'virtual reality'. And this world can metamorphose depending on which rival assumptions are held. We may assume that electricity would have been generated from the dirtiest coal plant currently in existence. But we may also assume that generation would have come from gas. Or other renewables. Or nuclear. But what is to stop us assuming that the new generation was simply not necessary in the first place due to demand management, or greater efficiency within existing power stations, in transmission and distribution or in end-use? It can be objected that some of these rival assumptions are more plausible than others. This is certainly true in the short-term, but once the time horizon is extended to five, ten or 20 years into the future, then an ever-wider range of options becomes plausible. Further, wind lobbyists usually do their calculations on 'avoidance' over a 25 year term – during which time many innovations can be expected. In consequence, the 'avoidance' term should not paint analysis into a corner. Quite different outcomes can legitimately be surmised on the basis of rival assumptions. This comment applies even more strongly in relation to emissions 'savings' and 'offsets'. With the latter, the inference that 'the electricity would have been generated anyway and by more polluting means' is more explicit. But this begs questions such as 'what exactly is being offset?' and 'what is the baseline?'. Close inspection of the usage of the terms usually reveals that no baseline is specified because the reality of what is offset is murky.

In summary, the terms 'cuts' and 'reductions' are best reserved for real and verifiable falls in emissions occurring in a recent past, as measured against a baseline. On the other hand, alertness is recommended towards usages of 'avoidance', 'savings' or 'offsets' which involve non-verifiable predictions, refer to an open-ended future and are made in the absence of a baseline.

some hours of the year, wind would be replacing other production forms, like gas' (Holttinen and Tuhkanen, 2004: 1639). This led them to predict a falling abatement curve to around 650g CO_2/kWh in scenarios of intermediate to large-scale wind penetration to 2010. On the other hand, in relation to the UK with its high penetration of gas in electricity generation, the Sustainable Development Commission (2005: 35) acknowledged that 'the actual CO_2 displacement in 2020 is hard to estimate and so for the purpose of this report, it has been assumed that wind output will displace the average emissions resulting from gas-fired plant. This figure is likely to be conservative, as in reality some coal-fired generation is likely to exist in 2020'. Hence there are grounds to consider that in the UK both coal and gas are displaced by wind, but uncertainty persists as to relative proportions. This is reflected in the 'standard' emissions displacement figure used by the

UK Carbon Trust of 430g CO_2/kWh. Meanwhile in France, with its 90 per cent GHG-free ESI, the current 'standard' emissions avoidance factor is 200–250g CO_2/kWh (Bataille and Birraux, 2006: 135). In summary, there are compelling reasons for considering that at the moment *some but not all* kilowatt hours of wind power replace kilowatt hours of coal-fired generation, with wide variation across national ESIs. This can only result in average emission factors lower than the 800–900g CO_2/kWh associated with displacement of coal. Finally, it might be supposed that empiric research has been conducted to produce reliable data on actual reductions and settle these uncertainties. In fact, at the time of writing no authoritative study could be found.

In the long term, the outlook is more open-ended. Around Europe, a number of aging coal-fired power stations will be closed during the period to 2020, but their replacements are not known. We may surmise that there will be some combination of wind, other renewables, gas, nuclear and advanced coal-fired plants (some with carbon capture and storage). All such replacements will have lower emissions than current coal-fired generation and so average emission factors are set to reduce. This means that the emissions displacement effect of a wind turbine built this decade, far from remaining constant, will fall more or less continuously throughout its lifetime. This puts into question claims made by wind lobbyists such as the following:

> on the assumption that the average value for carbon dioxide saved by switching to wind power is 600 tonnes per GWh, the annual saving under this scenario will be 1856 million tonnes of CO_2 by 2020 and 4800 million tonnes by 2040. The cumulative savings would be 11,768 million tonnes of CO_2 by 2020 and 86,469 by 2040 (EWEA and Greenpeace, 2002: 7).

There are, however, good reasons to think that most of the emissions assumed in this scenario for the distant future of 2040 will never happen. One such reason is that the average emission factor is already below 600 tonnes per GWh in several reference countries. Wind power cannot 'reduce' non-existent emissions. More recent declarations from the wind lobby have grown more restrained, with the GWEC and Greenpeace (2005: 174) admitting that 'specific avoidable emissions are going to decrease from 2000 to 2020'. Nevertheless, the presentation of emissions claims is an instance of how partisans of a technology use technical information 'in an advocacy fashion, that is, to buttress and support a predetermined position' (Sabatier and Jenkins-Smith, 1993: 218).

In summary, the investigation of wind energy as a means to displace fossil fuels and decrease GHG emissions has revealed a number of lacunae. The conceptualisation of causal pathways to emissions reduction effected by wind power lacks clarity – especially in media and 'public relations' presentations – leading to misunderstanding. Emissions avoidance is sometimes confused with emissions reduction. Circumstantial evidence exists to support the view that wind power usage *has* led to substantial GHG cuts, but no agreed methodology exists for their accurate calculation. Empiric research needs to be conducted into the displacement effects of wind on fossil fuel (and other) energy sources, within the day-to-day operation of national ESIs. On the basis of observation, a common methodology for the calculation of emissions savings can then be agreed and validated, such that reductions can be identified and quantified with confidence.

Greater clarity and accuracy matter, since the financial implications are significant. Current policies to promote RES-E do not contain a reward system for either emission cuts or 'avoidance', but this may change. Wind power is not eligible for 'carbon credits' *per se* – namely, payments per tonne of 'avoided' carbon – but its inclusion within schemes devolving from the Kyoto Protocol's 'flexible mechanisms' implicitly puts a price on carbon 'savings' achieved through its deployment. The wind lobby therefore has a material interest in maximising claimed carbon abatement, since the higher the 'avoidance' factor, the greater the potential for income generation. But if claims regarding carbon 'offsets' can vary by a factor of two (for example, 860 versus 430g CO_2/kWh), so too will the associated price spread. For example, the National Audit Office (2005: 36) estimated that carbon avoidance through the RO mechanism cost between £70 to £140 per tonne of CO_2. More troubling still, Ofgem (2007: 7–8) estimated the cost of carbon abatement through the RO as being in the range of £181–481 per tonne of carbon, as compared to £66/tC for the UK Emissions Trading Scheme, £18–40/tC for the Climate Change Levy and between £12/tC and £70/tC under the EU-ETS.[9] However, if the aim is to set a market price on carbon, as economists and policy-makers now argue, the current uncertainties over quantities and price spreads are too large. It is a truism of economic life that wrong price-signals lead to misallocation of resources. Carbon reductions can be achieved by a number of means, whose relative costs are being established. However, RES-E developers are accountable neither for their predictions on generation nor associated emission outcomes. In order to inform sound decision-making, it is crucial that the pricing of carbon *reduction* be reliable, consistent and lead to meaningful comparison across a range of options. Solutions to the methodology problems regarding emissions accountancy will be relevant in many contexts. They go well beyond wind power deployment in Europe, to consideration of its impacts in other parts of the world (through the application of the Kyoto Protocol's 'flexible mechanisms'), to analysis of the effects of other renewables, and to comparative evaluation of a variety of pathways to GHG cuts. Meanwhile, 'cap and trade' schemes such as EU-ETS increase the price of fossil fuel generation and help make wind power competitive without subsidy. For the future, it will be essential to clarify the current confused picture and to coordinate the spread of energy and climate policy measures in a coherent manner.

Conclusions

This chapter has situated the deployment of wind power within national ESI contexts in order to establish whether the sector has entered the mainstream of electricity generation. Rival conceptualisations of the role of wind power have been identified. Whereas enthusiasts sometimes believe it offers an independent source of supply, utilities have tended to view it as a mere 'fuel saver'. However, neither of these views captures the current trajectory of wind power. Wind energy remains an atypical source for generation, but no longer is it a minor one. Analysis has shown that its career is based on fuel displacement. The competitive threat it poses to other fuels is real and rising. Integration of wind power into national ESIs has required greater convergence with conventional generation. Ongoing innovations include revisions in the support system to meet the economic challenges of market integration and better management of production (through improved grid codes and centralised operation centres). But grid integration is problematic, partly because of the characteristics of wind power but partly due to grid design. Large-scale reconfiguration of the grid over the long term may resolve the difficulties, but will be costly. In combination, these innovations entail an increasingly centralised, 'bulk power' conceptualisation and deployment of wind power, which radically departs from the 'embedded generation' characteristics of its early days and pushes the sector onto a 'hard path'. Movement towards the mainstream heralds new patterns of industrialisation, a greater role for utilities, revised relationships with other parts of the ESI and domination of the wind sector by big capital. Ironically, mainstream integration involves abandoning the 'power to the community' tenets of wind pioneers.

At the same time, the composition of national ESIs is changing in ways that cannot yet be predicted. The future of both coal power and nuclear power hangs in the balance, pending technological progress, economic reassessment and shifts in political acceptability. But it is clear that wind power cannot replace both. Indeed, analysis of ESI operation has shown that the scenario of 'wind for nuclear' is not credible in practice, as wind power cannot provide base-load. Neither has the scenario merit on climate protection grounds, since – were it possible – it would merely substitute one GHG-free source for another. These two factors undermine the 'radical green' vision of a future ESI. The alternative scenario of 'wind for fossil fuels' is credible for the medium term, on both energy security and climate protection grounds. However, the accountancy of GHG emission reductions in the electricity sector must be improved. It is lamentable that, despite circumstantial evidence of emission cuts due to wind power, reliable data are not available. For the future, a variety of means will be required to reduce emissions from electricity generation, and it will be vital to identify the causal pathways to reductions with confidence and to quantify the economic and environmental outcomes of each with accuracy. Finally, the biggest challenges for the future lies in designing stable and secure ESIs (which retain diversity in energy sourcing whilst optimising complementarity between sources) and in balancing the institutions of 'state' and 'market' to achieve this aim.

7
Siting, Planning and Acceptability

Introduction

Having reviewed the integration of wind power into the electricity supply industry, we now turn to its integration within the physical environment. This chapter considers the acceptability of onshore wind power deployment from a number of viewpoints. It first looks at the siting question in terms of wind and land availability, and then analyses the extent and significance of the regional concentration of wind power in the five reference countries. The second section reviews planning processes and the criteria on which applications are decided, assessing the impacts of contrasting national planning traditions. A key issue is whether conflict resolution has been handled in a 'bottom-up or 'top-down' manner. Building on the discussion of these geographical and institutional dimensions, the third section draws lessons from cross-national comparison of factors which have favoured public involvement and acceptance of the technology, such as societal participation in strategic locational guidance, community ownership and wider socio-economic advantages.

Physical and geographical dimensions of wind power deployment

Siting issues

From the point of view of an investor or user looking to erect one or more wind turbines, the key factors determining site acceptability are wind speeds, the availability of suitable land and the scope for grid connection.

The availability of wind is the most crucial factor in site selection. The level of electricity generation depends on the prevailing wind regime, understood as the combination of average speeds and the number of hours per year of availability. Power extraction from wind is proportional to the *cube* of the wind speed. Thus an apparently small increase in speed leads to a large increase in output (see pp. 38–9 for more details). In general, wind speeds are higher at sea than on land, and in coastal rather than inland areas. In addition, wind speed increases with height. The size factor has pushed wind power technology towards the building of ever-larger towers and rotors. Multi-megawatt machines provide greater performance and profitability, especially as their 'footprint' is similar to smaller ones, minimising ground rent. But the height factor has also made deployment on hilltops economically attractive. The scope for land use conflicts arising from these factors was already identified in the early days of wind power, as the following quotation illustrates:

> to benefit fully from the cube law relationship between wind velocity and power output, the best sites on land are usually hilltops or other exposed sites. In the UK it seems unlikely that a very large number of wind turbines could be sited on hilltops because many of these are in areas of outstanding natural beauty (Taylor, 1983: 12).

Thus wind availability rapidly shades into land availability and the criteria by which the use of land for construction of turbines is deemed appropriate.

The developer's perspective on land availability is determined by economic, physical and planning issues. Land must be available at appropriate rents, but site access is paramount. The erection of wind turbines requires extensive infrastructure. Transportation of tower components, nacelles and associated hardware – such as tall cranes – requires some of the largest articulated lorries currently using the road system. Finding ways to move these wide loads through narrow villages without modification to roads or buildings can be insuperable, and the problems increase once on-site, since lorries often ascend hilly terrain on 'temporary' tracks. Thus the physical characteristics of some potential sites rule them out altogether, or lead to compromises in terms of installing smaller turbines and/or smaller arrays. Crucially, the developer needs to anticipate whether a substantial conflict of planning priorities would rule out gaining a consent, raising considerations related to land-use designation, environmental desiderata and social preferences to which we return below.

Land availability also needs to be considered from the societal point of view. Most renewable sources derive directly or indirectly from solar energy and have a common drawback: 'meeting large-scale energy needs with solar energy requires large collection areas because of its low concentration' (Bent, Bacher and Thomas, 2002: 36). Usable wind – as distinct from gales, hurricanes and tornados – has low energy intensity, with the consequence that wind farms require extensive open areas in rural settings (or else offshore). Urban areas are much less suitable because of lack of open spaces and turbulence in the wind caused by buildings, which reduce its usability. Further, as noted by Everett and Boyle (2004: 386): 'most of our energy demand actually occurs within the relatively small areas of our major cities. Many of these are built in sheltered inland valleys, deliberately to avoid the worst excesses of wind and wave energy'. In consequence, wind farms are typically sited at considerable distance to the built-up areas which constitute the load centres. Thus land availability for wind power correlates with population density, with lower density tending to favour deployment. But rural siting poses the technical questions of power transport and grid configuration discussed in Chapter 6. Rural communities are not used to the presence of large-scale electricity generation facilities, nor do they necessarily draw benefits from the wind farms on their door-step. On the other hand, the urban consumers of large-scale RES-E – as distinct from small-scale or micro generation – are at a physical remove from rural sites of generation, and often lack understanding or experience of the technologies deployed for their benefit.

Regional concentration

Desiderata arising from wind and land availability have led to the regional concentration of wind power in all of the reference countries, with social, environmental and technical consequences.

Denmark is well-endowed in terms of wind conditions, having long coast lines and mainly flat landscapes. Offering few obstacles to winds from the North Sea, the western coastal areas of Jutland – especially the least inhabited northwest corner – has seen most wind deployment. In 2003, western Denmark had a total of 2400 MW, but only 600 MW in the east (Eriksen and Hilger, 2005: 199). Given that Denmark has separate grids for west and east, wind power concentration has led to significant load management problems in the western sector, as well as problems of onshore saturation.

Germany, with its continental land-mass and limited coastline, has middling wind conditions. Its geography has had a significant impact on wind deployment because of differences in wind speeds, with around 7 m/s at coastal sites to well below 6 m/s inland (Lehmann, 2003: 40). Most development has been in Schleswig-Holstein, Lower Saxony and Mecklenburg-Vorpommern, which are northern coastal *Länder*. Aided by their proactive policies, these three states accounted in 2005 for 8274 MW of a total 18,427 MW of wind capacity in Germany (DEWI, 2006). In addition, Brandenburg, Saxony-Anhalt and North Rhine-Westphalia – also situated in the north – accommodated 7037 MW. Deployment in the lower wind states of southern Germany has been very limited. In 2005, Bavaria and Baden-Württemberg together had only 520 MW (DEWI, 2006). This marked north-south split results in the grid problems discussed in Chapter 6. But the regional distribution of wind deployment is unlikely to change significantly, since the 2004 EEG amendment curtailed entitlement to feed-in tariffs in low wind speed areas. Limited availability of sites has led to a peak in growth rates. The Federal Environment Ministry has stated that 'the potential for wind energy utilisation at onshore windy sites is already largely exhausted' (BMU, 2004a: 9). New sites are still being found, but are in short supply.

Spain has seen a more balanced pattern of deployment due to geographical and demographic conditions. Not only does Spain have a large land mass, it is a peninsula benefiting from long coastlines, hilly regions and low population density in its many rural areas. Both wind and land availability are therefore high. Wind farms are built in sparsely populated countryside, often in mountainous areas. Early experiments were in the south, in Andalusia (notably Tarifa). But the pioneers for large-scale deployment were in the north, with Navarra – one of the smallest but least populated regions – taking an early proactive stance (Garrigues, 2002: 149). Turbines were often placed on hard-to-access hill-crests, leading to use of smaller turbines than in northern Europe (Avia Avanda and Cruz Cruz, 2000: 40). The process resulted in a major transformation of the landscape in northern Spain, for example along the *Camino de Santiago* (St James' Way). More recently, Navarra has been overtaken by neighbouring and central regions. In 2005, Galicia had 2371 MW of capacity, Castilla La Mancha had 1871 MW, Castilla-Leon 1759 MW, Aragón 1333 MW whilst Navarra stood at 908 MW; together these five regions had 8242 MW of Spain's total capacity of 9781 MW (IDAE, 2006: 9).

These data point, therefore, to a measure of regional concentration. Even in 2005, Extremadura and the Madrid area had no wind farms

(AEE, 2006: 45). Although the poorest region in Spain, Extremadura banned wind farms on landscape grounds and to protect rural tourism, but in 2005 the regional authority reversed its policy on financial grounds (Anon, 2006). Over time, dispersal of wind farms in Spain has affected more regions and landscape types, with a greater number now being located in low-lying, flat terrain. Availability of land for further wind deployment is still high in Spain – as is the appetite for construction which marks the country – but grid weaknesses are an obstacle.

In continental France, Brittany and the north coast together with Languedoc-Roussillon (bordering the Mediterranean) have the best wind regimes, with average speeds of 6.5–7.5 m/s. They have also seen the largest concentration of wind power. In 2003, 44 per cent of France's 218 MW were in Languedoc-Roussillon (Chabot, 2003). But by the end of 2005 the region had only 18.3 per cent (138 MW) of France's 757 MW (Chabot, 2005). A dispersal of new build to other regions is thus in evidence, given that a broad swathe of inland sites have wind speeds in the 6.5 m/s range. Geographical patterns of wind deployment are in flux, as France is a relative newcomer to wind power.

The UK is renowned for being an island system with the best wind regime in Europe. Winds are generally stronger in the west, with north-west England and Scotland having outstanding availability. The first wind farm was built in Delabole, Cornwall by a local farmer in 1991,[1] very much following the Danish model. Thereafter, however, wind farms have – with few exceptions – been commercial developments (rather than individual or cooperative initiatives). Capacity growth in the 1990s was dispersed around the UK, with some concentration in Welsh uplands. In the 2000s, Scotland became the developers' area of choice due to excellent wind conditions, low population density and the proactive stance of the Scottish Executive. In November 2006, onshore operational capacity stood at 934 MW in Scotland, 301 MW in Wales, 291 MW in England and 112 MW in N. Ireland (BWEA, 2006).

In summary, wind and land availability are – unsurprisingly – the main reasons explaining the skewed geographical distribution of wind power. Yet political, economic, social and environmental factors have also contributed to regional concentration. A fine-grain analysis of patterns of land-use indicates differences in deployment patterns in terms of landscape types and land ownership. Whilst specialists make subtle but important differentiations between landscape types, the major choice that must be flagged in relation to wind power is the use of upland over and against lowland sites. This alternative can also overlap with the dimension of ownership. In Denmark and Germany agricultural plains owned by farmers are the main locational choice. Being under intensive agriculture, the ecological value of this land is considerably reduced. Farmers and owners are concerned with economic livelihood and diversification of incomes in a sector that has struggled in recent years. The picture in France is evolving, but the scope to use farmland is high in what is an extensive agricultural country. In the UK however – and to an extent in Spain – the tendency has been to build wind farms on upland sites. These sites are often ascribed high landscape value, set aside from the mainstream of economic exploitation and, in some cases, are common land. Further, England is a small country with a very high population density, especially in the south. Wales and Scotland have lower population densities and retain areas characterised by 'wilderness' or 'wildness'. Although these terms are contestable, the rarity of 'unspoilt' countryside in the UK is commonly acknowledged, as is the need to preserve what remains. A crucial difference between lowland agricultural land and upland commons is that the former category has significant pre-existing infrastructure (facilitating access and construction), whilst the latter may have none. Where wind farms are built in 'wild' or 'unspoilt' areas, roads and tracks have to be constructed, changing the accessibility, usage and character of the landscape. Thus the impacts of wind farm development are of a different order of magnitude in the two categories. Uneven impacts signal the complexities and sensitivities which surround planning processes.

Institutional dimensions: the planning process

Planning and permitting involve setting criteria of acceptability and applying them to particular projects. Conflicts of perspectives, values and interests arise which must be settled in the planning process, often on the basis of imperfect compromises. As observed by Cullingworth and Nadin (2002: 2): 'if there were no conflicts, there would be no need for planning. Indeed, planning might usefully be defined as the process by which government resolves disputes about land uses'. Inevitably, planning criteria change over time, more or less in line with changing land-uses and with changes in values and interests. Reforms in criteria form part of a policy learning process, with the learning curve being particularly marked in relation to new technologies. Wind power has posed unique planning challenges due to the height of turbines, their spread across the countryside and their intrusive infrastructure which have effects *on the ground* with roads and tracks for articulated vehicles, construction of substations, buildings and so forth, effects *underground* due to excavations for deep foundations and buried cabling, and effects *overground* from large towers and moving rotors plus overhead wires and pylons. Due to this scale of impact, a controversy has arisen over whether wind power complements existing rural land-uses or constitutes an industrialisation of the countryside. To deal with these issues, wind farm planning has involved a fraught social learning process over the long term.

A frequent starting point for determination of planning applications is that proposed by the ODPM (2004b: 167): 'the material question is whether the proposal would have a detrimental effect on the locality generally, and on amenities that ought, in the public interest, to be protected'. For wind power, this general and abstract consideration is translated into operational criteria, including proximity to dwellings and other buildings, electromagnetic interference (with effects on civil and military transmissions such as television and radar), landscape designations and so forth. An important procedural element within EU member states is use of environmental impact assessment (EIA), set up under European directive 85/337/EC as amended by directive 97/11/EC. The EIA procedure is required for all but the smallest wind installations and is undertaken by the developer. It is meant to provide an objective report on a proposal's effects, *inter alia* on ecology (wildlife and habitats, soil, water, air), human beings and material assets (including noise and nuisance), cultural heritage, landscape impact and visual impact. Although a number of common problems have thereby found common solutions, national planning traditions continue to mark practice and produce divergence. Decisions on permitting are informed by national, regional and local planning guidelines, whose form and content vary by locality and over time. Because of the detailed nature of planning determinations, the following discussion makes no claim to exhaustive treatment, but aims only to identify key characteristics and staging posts in the evolution of national planning practices related to wind power.

Denmark

Denmark as the wind power pioneer was the first to encounter planning problems and the first to find solutions, many of which have been replicated elsewhere. As acknowledged by the Danish Energy Agency (1999: 14) 'the first generations of wind turbines were often sited with no official planning and zoning restrictions'. This included sites where development is now controlled, such as nature conservation areas and near dwellings. The combination of proximity to housing and noisy turbines tarnished the reputation of wind power and provoked social opposition. When in 1992 a national directive on wind power deployment was proposed in the *Folketing* (Danish Parliament), a clear majority opposed it (Van Est, 1999: 101). Recognising that 'use of renewable energy sources will, however, have negative consequences on the environment, in the form of effects on the landscape and nature' (Danish Ministry of Environment and Energy, 1996: 75), the authorities learned to take a 'softly softly' approach to planning reform.

Subsequent to the work of a 'Wind Turbine Siting Committee' set up in 1991, the Danish government in 1994 issued a circular on local wind power planning and by 1997, 205 out of the 273 municipalities had responded, identifying a total of 5065 sites for a capacity of 2381 MW (Hvidtfelt Nielsen, 2005: 116). The procedure was decentralised in that no quotas were set by national government. Rather the Ministry of Environment and Energy coordinated national planning, regional planning in the counties and local planning in the municipalities, for example in relation to high-voltage transmission lines in rural areas (Danish Energy Agency, 1999: 13). Further, Denmark has an 'Act on Spatial Planning' empowering county authorities to prepare regional planning guidance, whilst municipalities are the local planning authorities which specify wind power development zones. The core characteristic of 'spatial planning' as understood in this context is the creation of 'inclusion' and 'exclusion' zones, which provides strategic locational guidance to inform planning decisions. Clear statements on the suitability (or otherwise) of locations for particular purposes offer a relatively high level of predictability in relation to planning outcomes. In keeping with a 'bottom up' approach, the designation of zones was undertaken through a consultative process involving public hearings. This participative procedure was a contributory factor to gaining local acceptance.

During the period of planning reform, the pace of new wind power installation slowed down (Danielsen, 1995: 61; Toke, 2002: 95). Subsequently, record growth in 1999 and 2000 meant that the target of 1500 MW for 2005, which was set in 1996,[2] was exceeded several years early. In a context of unexpectedly rapid expansion, the planning authorities showed reluctance to designate new sites (DWIA, 2002: 10). The stalemate has persisted, since aggregate capacity of 3128 MW in 2005 was well ahead of target, whilst subsidies to new build had largely been withdrawn. Although the 2006 'repowering' target of 350 MW will mean fewer old turbines, the larger size of new machines necessitates revision of zoning, and a national planning committee was established to improve long-term planning (DWIA/Holst, 2006).

Over time, prescriptions governing turbine siting have been tightened and systematised. Local plans specify height, colour and construction of towers. Noise levels are set nationally, with limits of 45 dB(A) for dwellings in rural areas and 40 dB(A) in residential areas (Danish Energy Agency, 1999: 14). Minimum separation distances are prescribed for housing, monuments and churches, and for coastlines, lakes, streams and forests. Designated zones avoid conservation areas and prized landscapes. The Danes also pioneered the aesthetics of a spatial ordering of turbines. The wind farm concept was founded on the norm of geometrical design. To achieve integration into flat, agricultural landscapes, attention is paid to linear alignment and equal spacing (Nielsen, 2002: 115). In principle, cumulative effect is avoided by clear separation between groups of turbines. Moreover, Denmark has no large wind power stations. Early installations were often solitary turbines, or small clusters. Even in 2005 the largest onshore wind farm was the Rejsby Hede installation of 42 turbines with a combined capacity of 21 MW – which was built in 1995 (IEA, 2005a: 88).

Finally, ownership restrictions were an important but frequently overlooked component of Danish planning. Individuals (including farmers) were only allowed one grid-connected turbine on their own land (Danish Energy Agency, 1999: 11). Wind turbine investors – whether individually or through cooperative schemes – had to live in the same municipality as the site (or neighbouring on it), and investments were capped in relation to specified consumption levels (Tranæs, 1996; Olesen; Maegaard and Kruse, 2003: 3). This was in order to avoid speculation in wind energy, which was considered to be undesirable (Frandsen and Andensen, 1996: 850; Hvidtfelt Nielsen, 2005: 111–12). Because of residency requirements, 'big, absentee investors are kept out' (Nielsen, 2002: 128). Indeed, in the early 1990s the utilities ran into strong local opposition when seeking planning permission for their wind farms (Christensen and Lund, 1998: 3). Although residency restrictions were lifted in 2001, they help explain the absence of large wind power stations. Whilst locally-owned schemes are often said to be *preferable* to achieve success in gaining a planning consent, during the heyday of wind power expansion in Denmark local ownership was generally *compulsory*.

In summary, the Danish spatial planning approach fostered a 'social contract' involving the expansion of wind power through predictable and orderly development. Whilst no planning regime can eliminate controversy, the Danish system struck a balance in terms of who gained and who lost. It was characterised by realism in its recognition of conflicts of interest and by its attention to resolving them through a significant degree of 'bottom-up' participation, with public involvement leading to community acceptance.

Germany

A planning model similar to Denmark's was used in Germany, again with considerable success. National planning reform favourable to wind power was undertaken. The Federal Building Code specifying which structures were permissible in the countryside was revised in 1996 to include wind turbines (Rickerson, 2002), thereby giving them the same priority in permitting processes as conventional power stations. Since Germany has a federal system, decision-making takes place within the *Land*. This reduces the felt presence of central government. The major output of the German spatial planning regime is the regional development plan.[3] Local authorities designate, on a 'bottom-up' basis, areas deemed suitable or unsuitable for wind power deployment. The presumption in favour of wind farms within 'priority' zones eased the process of spatial integration. Conservation interests are protected as national parks, bird sanctuaries and other designated areas are – in principle – excluded. Nevertheless, zoning measures have shown their limitations. Given the north-south split in wind availability in Germany, the northern *Länder* have drawn up clear provisions for wind farm zones, but the authorities in some southern *Länder* have shown reluctance. Yet even in the north, planners have found it difficult to keep up with the pace of wind deployment. A temporary moratorium was declared in Schleswig-Holstein to allow for broad-based consultation over 1997–8. This allowed identification of 166 suitable zones, whilst banning wind farms from elsewhere in the territory. In Brandenburg, a largely unregulated boom followed the installation of the first wind turbines in 1993. A regional development plan for the picturesque Uckermark (part of Brandenburg), identifying 26 'special wind areas', was completed only in 2000. However, successive wind booms have outstripped provision of designated sites in some areas.

As in Denmark, regulatory review clarified and improved construction requirements related to turbine dimensions, separation distances and so forth. An unpopular requirement, however, is for turbine blades to display flashing red lights (as an air safety measure). With wind farms being now constructed in close proximity, the issue of 'cumulative effect' has to be addressed in the environmental impact assessment. In summary, the concentration of wind farms in northern Germany indicates the success of the planning regime, a supportive political environment and the impact of feed-in tariffs, but is currently raising questions of land saturation and public fatigue.

Spain

The 'autonomous' Spanish regions enjoy considerable powers of decision-making, *inter alia* having competences in energy policy (Faulin *et al.*, 2006: 3). Planning permits are issued by the regional authorities, whose energy departments decide approval terms and procedures for RES-E plants under 50 MW (Dinica, 2002: 215). Spatial planning is done at the regional level. In relation to the 'zoning' phase, Blázquez, Calero de Hoces and Lehtinen (2003: 468) noted that 'the example of Tarifa gave an impetus to the now common procedure of negotiating territorial plans for the organisation of wind resources'. The 'autonomous' governments took a proactive stance towards the promotion of wind power, and regional targets were established by the late 1990s aggregating to 10,000 MW for the next decade (Avia Avanda and Cruz Cruz, 2000: 38). Social, economic and environmental stakeholders bargain over the designation of inclusion/exclusion boundaries. Firms are granted concessions in specified 'areas of investigation' (EWEA and Greenpeace, 2002: 21).

Notwithstanding the rapid pace of wind power deployment, Spanish administrative procedures are complex and demanding. Río and Gual (2006) reported practitioner claims that 'bureaucracy often makes administrative procedures a lengthy nightmare, with regions following their own singular procedure'. Aubrey (2005: 20) referred to an 'administrative labyrinth' involving 60 regulations and 40 categories of proceedings at national, regional and local levels, such that large projects take over five years to come to fruition. Considerable variation appears to exist regarding the content of regulatory measures and the effectiveness of their application.[4] Hurtado *et al.* (2004: 484) noted the existence of a national law on noise as well as regional regulations concerning the impact of wind farms on nature conservation, but there is no norm regarding visual impact. The handling of environmental impacts has been uneven, apparently with a greater measure of success in Navarra than in Galicia and Castilla which have seen conflicts with organisations and residents (EWEA and Greenpeace, 2002: 21).

The Spanish regions continue to promote wind power actively. In the 2004–5 planning round, proposals from the regional authorities aggregated to some 40,000 MW and were around double the targets contained in the PER, the national planning document (AEE, 2006: 30 and 53). Indeed, regional authorities have been so inundated with wind farm proposals that several – Aragón, Navarra, Asturias – temporarily suspended new applications, due to excess demand and grid saturation (AEE, 2006: 21). The government and the grid operator, REE, are analysing the outcomes of the regional RES-E deployment plans to assess compatibility between requests for connections and grid capacity. Thus saturation issues arise in relation to the grid, but not due to land availability or lack of political enthusiasm.

United Kingdom

The place allotted in the UK planning system to development plans does not stretch to zoning measures of the Danish, German or Spanish varieties. Instead, the UK tradition stresses 'criteria-based' decision-making. This means that 'the British system embraces discretion and general planning principles, rather than certainty for the landowner and developer' (Cullingworth and Nadin, 2002: 49). Acceptability criteria to assess a planning application include conformity with development plans and policies, due regard for landscape designations, as well as positive and negative impacts on the locality (ODPM: 2004b: 58). However, development plans do not typically specify inclusion/exclusion zones *per se*. Planning permission for electricity generation plants of under 50 MW nominal capacity is granted by the local planning authority under the 1990 Town and Country Planning Act but, under section 36 of the 1989 Electricity Act, the final decision for larger plants lies in England and Wales with the Secretary of State for Energy and in Scotland with the Scottish Executive, informed by a public inquiry where the local authority has refused consent. Thus the UK permitting system for large generation plants includes a significant measure of centralisation. This is relevant to wind power deployment, since the trend towards large wind power stations has meant that central – rather than local – authorities often decide planning applications.

It is sometimes mooted that approval procedures for wind farms are uniformly cumbersome and slow, and the climate in which they are conducted is unfavourable. Survey results from the National Audit Office (2005: 22) revealed a more complex picture:

Between 1999 and 2003, 94 per cent of planning applications in Scotland received approval against 50 per cent in England and

40 per cent in Wales, although this latter figure masks a significant legacy of wind farms in Wales that were approved before 1999. Within England, between 1999 and 2003, local planning authority approval rates for wind farms ranged from zero to nearly 100 per cent, with most authorities reporting approval rates of between 50 and 80 per cent. As regards the average time for applications to go through the planning process, for England this is 8.5 months, for Scotland 10.0 months, and for Wales 23.4 months. There is also significant variation at a regional level.

Development in Northern Ireland has so far been limited, whilst the low percentage of approvals in Wales has to be taken in the context of the 'wind rush' of the late 1990s which led to a backlash against wind power. The middling percentage of approvals in England was explained by Toke (2005b: 1530) on the basis of 'three (independent) variables, namely the opinion of the parish council, the planning officer's recommendation, and the opinion of the countryside protection group, are strongly associated with the decision of the local planning authority (the dependent variable)'. The parish council is the body closest to and perhaps most representative of the residents living nearest to a proposed development and most affected by it. The influence of countryside protection groups in the UK planning system is long established. Cullingworth and Nadin (2002: 10) commented that 'of the many important factors which have shaped the system are the strong land preservation ethic, epitomised in the work of the Council for the Protection of Rural England (and its Scottish and Welsh counterparts)'. These groups have long sought to minimise intrusive landscape and visual impacts in unspoilt areas. The very high proportion of approvals in Scotland is partly explained by the year 2000 reform of planning guidelines, with NPPG6 being the relevant document for renewables.[5] Of equal if not greater importance, however, are broader institutional and political features. Scotland has seen many applications for wind power stations above 50 MW. Under section 36 of the 1989 Electricity Act, these are referred to the Scottish Executive which, on the basis of a political decision to promote renewables, approved all bar two up to 2005.

The high success rate in Scotland as compared to variable outcomes elsewhere encouraged interested parties such as the BWEA to press for planning reforms in England and Wales. Guidelines for renewables contained in PPG22 dated from 1993,[6] and were considered out-dated and less supportive than NPPG6. The 2001 introduction of the RO stimulated large numbers of wind farm planning applications, with the consequence that many local authorities had to take decisions on wind power for the first time. Lack of experience on the part of the planners contributed to logjams and uneven outcomes. Revised national documents were seen as the means to provide guidance on new development control issues and to produce speedier and more consistent results. In 2004, the Office of the Deputy Prime Minister circulated a new 'Planning Policy Statement' applicable to renewables, numbered PPS22. Disseminated with a clear intent to favour the expansion of renewables and accelerate consent procedures, PPS22 specified three elements to be taken into consideration in planning decisions: targets, criteria-based policies and locational considerations (ODPM: 2004b: 21). The reform represented a very British compromise. The emphasis on criteria-based decision-making retained the main characteristic of the UK planning system, with its scope for local discretion. On the other hand, the New Labour stress on targets introduced top-down requirements. The reference to 'locational considerations' provided a fudge. Official guidance glossed this concept in terms of 'broad areas' but 'without definable boundaries' (ODPM: 2004b: 22), and so seemed to rule out designation of inclusion and exclusion zones on the continental model. But in practice, sub-national contrasts persisted. Thus whereas 'spatial planning' of wind power is avoided in England, in Wales a system of 'areas of search' was brought forward in the TAN8 planning document, under the devolved powers of the Welsh Assembly.

The consequences of the reform were still working through in 2006–7. BWEA (2006) data suggested that UK wind power capacity would soon double. Onshore, 1639 MW were in operation and 590 MW in construction, whilst a further 1535 MW had been consented. Offshore, 303 MW were in operation and 90 MW in construction, with 765 MW consented. Projects in planning came to 4283 MW offshore (with an average project size of 428 MW) and 7653 MW onshore. Of the latter, 5316 MW were in Scotland, with an average project size of 70 MW. Thus whereas in 2006 Scotland had 57 per cent of UK operational onshore wind capacity, that proportion could rise dramatically. Scotland regularly beats the record for the largest wind farm in the UK (and possibly Europe): in 2005, Blacklaw at 97 MW; in 2006, Hadyard Hill at 120 MW; and Whitelee at 322 MW was consented in 2006.

The complexities of the UK planning system with its sub-national variations should not distract from the three main outcomes for wind power in the mid-2000s. First and foremost, the scale of new wind power capacity being consented and built is of a different order of magnitude as compared to the 1990s. Secondly, most new capacity is achieved through large-scale installations, both on and offshore, with Scotland being the developers' preferred choice. Thirdly, under section 36 of the 1989 Electricity Act, wind power stations above 50 MW are consented by central – and not local – government. Large-scale build in the maritime environment no doubt requires a high level of national government coordination. However, the onshore Scottish case presents an oddity in that decision-making on wind power stations is devolved from Whitehall to the Scottish Executive – but taken away from local authorities. As developers increasingly opt for large installations requiring national government determination, the question of accountability is raised within the planning system. Whilst a logjam in relation to the 5316 MW in planning in Scotland can be adduced to administrative overload, it may also signal heightened political sensitivity towards a potential backlash caused by rapid transformation of Scotland's rural environment. Thus the 'top-down' approach is effective so long as it can either sweep resistance aside or carry the conviction of the community.

France

The French approach to planning is similar to the UK in being largely discretionary, criteria-based and decided on a case-by-case basis. At the same time, although a spatial planning approach comparable to Denmark and Germany is not strongly in evidence, neither is there the same strong resistance to it found in England. France's intermediate positioning explains similarities in wind power planning outcomes with the UK, but it also offers more scope for switching to spatial planning.

As in England and Wales, the slow rate of build in France has been attributed to delays in obtaining construction permits (Chabot, 2005: 6). Decision-making is local in each case, though the institutional frameworks are different. In France, the mayor of the municipality will vet a planning application – and can effectively veto it – but consent is given by the department prefect (*préfet de département*). For approval of a wind farm, some 27 clearances from several independent public bodies are required (Boston Consulting Group, 2004: 22). As in the UK, delays have arisen from the complexity of procedures, lack of planning experience with wind farms in local administrations and logjams due to the surge in applications after announcement of a new subsidy. In 2004–5, a decision on a wind farm application took eight months on average, whereas the target period has been five months (Ministère de l'Economie, des Finances et de l'Industrie, 2005b). In the interim, developers can be disadvantaged by a reduction in the feed-in tariff, or lose out on a grid connection. Nevertheless, official data on wind farm planning applications indicate that 65–70 per cent prove successful, with 852 MW approved in 2004, and 1557 MW in 2005 (Ministère de l'Economie, des Finances et de l'Industrie, 2005b). The quantity of approvals is considerably higher than the recent build rate, and this together with applications in process (which stood at 3198 MW in 2005) suggest that on current trends well in excess of 4000 MW of capacity will go on-line this decade.

However, this would still leave France short of the 21 per cent RES-E target by 2010. Efforts have been made to reform the planning system so as to reduce uncertainties and speed up the process. This would reduce the transaction costs of planning for both applicants and administrations. The 2005 French Energy Bill made provision for so-called 'wind power development zones'. Local authorities will designate these zones on a 'bottom up' basis and communicate plans to the prefect for approval. The designation process will involve setting of minimum and maximum capacities by zone, whilst taking into account landscape and nature conservation, protection of heritage sites and availability of grid connections. At the time of writing, the process is underway but zones have yet to be finalised. Inclusion zones will – in principle – allow fast-tracking of applications, whilst exclusion zones avoid wasted effort. The proposals are similar to the German and Danish models where zoning measures, together with a feed-in tariff and local ownership, have been the main drivers in wind power expansion. The French case therefore indicates an interesting example of policy learning by diffusion.

Synthetic overview

Reporting on expert opinion on best practice in wind farm planning, the IEA (2005a: 20) concluded that 'hierarchical planning is destructive' whilst 'collaborative planning is crucial'. Although a bold generalisation, it receives qualified support from comparison across the five reference nations. The policy learning process has passed through three development phases in the northern pioneer countries:

1. early shortcomings due to lack of adequate procedures, leading to problematic and sometimes illegal build;
2. improved development control with EIA procedures and 'bottom up' local/regional zoning measures, together with greater attention to conflict management;
3. recent rapid build, leading to onshore saturation and various degrees of community fatigue or opposition (discussed in the next chapter).

The Spanish case has passed through the first two phases, but with a large landmass and low population density, shortage of land is not a problem whilst appetite for further build remains high in many 'autonomous' regions. France, as a latecomer known for its high level of regulation, has seen few phase one type difficulties and has moved to phase two through the designation of 'wind power development zones'. As noted by Lake (1993), zoning is a means adopted by the state to reduce uncertainty for land developers. Lower planning risk ought to depress transaction costs and reduce the need for subsidy.

The UK has seen a complex pattern of wind power deployment because of its composite nature, its distinctive planning traditions and the availability of the wind resource. With limited exceptions, the planning process relies on a criteria-based style of decision-making, involving discretion and significant uncertainty over outcomes. Planning reform instigated by the ODPM in the early 2000s – of which PPS22 is only one manifestation among several – may, however, be going in a more 'top-down' direction with greater centralised control, for example through the setting of regional wind power targets. Thus the UK differs from the continental cases in two components of its planning process, namely decision-making style and the locus of control. These institutional characteristics help account for the significant friction existing between parties in UK wind power planning. Due to its aversion to the continental solution of 'bottom-up' zoning, the UK has largely skipped the first two development phases and moved directly to phase three – rapid build, saturation and social opposition (discussed in the next chapter).

Public involvement and acceptance

Siting and planning are not limited to technical and administrative issues. They also reveal that the involvement of the public – and especially the local community – is paramount. A high level of social acceptance can arise from three causes: (1) societal participation in strategic locational guidance; (2) community ownership and (3) wider socio-economic advantages. Examples of each category will be taken from the national case-studies in order to understand the dynamics of social acceptance as *process*, and not just as *product*.

Societal participation in strategic locational guidance

Denmark is a small country with a long tradition of democracy. Associated with its strong democratic values is a mode of societal problem-solving based on open dialogue and consensus-seeking. The first step to resolving disagreement is usually to acknowledge the existence of a problem. From the early days of wind power, the Danish government recognised the problems associated with the siting of wind turbines – instead of brushing them under the carpet as sometimes happens in other countries. Thus the Danish Energy Agency (1999: 8–9) acknowledged that 'the environmental advantages of wind power are on the global or national level, whereas its environmental disadvantages are on the local or neighbourhood level. (...) Such local disadvantages can lead to a lack of public acceptance of wind farms'. This frank recognition was a precondition for addressing problems associated with wind power deployment.

The will to anticipate and avoid social conflict was evidenced at various levels of civil society. The reconciliation of nature protection and wind power deployment required deliberate strategies (Christensen and Lund, 1998). Hvelplund (2002: 67–8) noted that alternative energy NGOs formed 'a collaboration with the established and large environmental protection organisation (*Danmarks Naturfredningsforening*) which made it possible to avoid an early stage confrontation between the "landscape" interests and wind power interests'. In contrast, 'landscape' and 'environmentalist' NGOs in the UK have been unable to collaborate.

Danish spatial planning procedures included an important participative element, involving local hearings. Whilst pressures arose to designate inclusion zones for wind turbines, spatial planning – in Denmark and elsewhere – also allowed the designation of exclusion zones. This is an important safety valve for the release of social anxieties. It contrasts with the UK discretionary system where uncertainty is ubiquitous, even in relation to the protection afforded by significant landscape designations.[7] Thus although the extent of local community involvement in the 'preplanning' phase varied in practice, 'bottom up' participation on an institutionalised basis contributed to Danish public acceptance (Sørensen and Hansen, 2001: 31).

Community ownership

The positive contribution of community ownership to social acceptance in northern Europe has often been stressed and was summarised by Frede Hvelplund as follows:

> One of the main historical secrets behind the Danish wind power success was that a system of cooperatives and neighbour and local

ownership was furthered by public regulation, resulting in more than 120,000 wind turbine owners in Denmark. People seem to like wind turbines, when they own them, and are not annoyed by the noise and visual inconveniences, especially when receiving a fair compensation. However, with a system of distant utility or shareholder ownership, the local inhabitants are only getting the disadvantages without any compensation. This is seen as unjust and results in increasing local political resistance against wind power. (Hvelplund, 2005: 237)

Community ownership had two enabling dimensions: as *opportunity* and as *reassurance*. Investment in wind power represented an *opportunity* – to generate and use electricity on-site, and make money by sales to the grid. Field-work contacts made clear that for many Danes individual or cooperative ownership represents a kind of 'pension fund' – a reliable if unspectacular stream of income over the long term. Residency requirements for ownership reinforced this approach, since investments were capped. Most importantly, it meant they were local. Residency controls offered a *reassurance* regarding the equitable distribution of benefits and burdens, since those who suffered nuisance were largely those that benefited. They also offered legitimisation for repelling large or distant investors, who might profiteer from public subsidies. Consequently when the utilities in the 1990s sought to build wind farms, they met resistance. Energy Minister Bilgrav-Nielsen acknowledged that the cause for utility difficulties in gaining planning consents was 'largely because of opposition from the Danish public which, whilst supporting wind power development by private groups of individuals, does not welcome the idea of sizeable utility projects being forced upon it' (quoted in Van Est, 1999: 101). Further, the authorities accepted this social preference, with the consequence that wind power stations have not been built by Danish utilities onshore – they have been relegated offshore. In Germany too, local and cooperative ownership has been important in promoting social acceptance. But the favourable dynamic created in Denmark by combining *opportunity* with *reassurance* has proved elusive elsewhere. Nevertheless, the stress on socio-economic opportunities has undoubtedly contributed to acceptance in other countries too.

Wider socio-economic advantages

A number of economic benefits have supported the development of wind power in the pioneer countries. Foremost among these is wealth and jobs creation.[8] In its early period, the location of the wind turbine industry overlapped its geographical area of deployment. The world number one, Vestas, originated in Jutland which is where the largest concentration of wind installations in Denmark is found. Likewise northern Germany was home to several major turbine manufacturers. In Spain, regional government policy contributed to firm creation. A key example was Navarra, where the authorities took a proactive policy in setting up EHN, a renewables based generating company, contributing 38 per cent of its capital and a further 37 per cent coming from Iberdrola (Garrigues, 2002: 149). Thus in the 'pioneer' countries the symmetry between the area of manufacture and the area of installation meant that the benefits of wealth and jobs creation were felt locally and regionally.

However, as the wind industry extended, this 'organic' overlap reduced. In some cases, it has been superseded by political insistence on local content requirements. In its 2006 wind farm tendering process, China not only prescribed 100 per cent local manufacture, but awarded contracts exclusively to Chinese state-owned firms (Jianxiang and Knight, 2006). Although this is an extreme case, precedents exist in Europe. In Spain, regional wind power targets equated to industrial plans, since they promoted not just electricity but also economic development. Regional governments tied planning approvals to inward investment, construction of factories and job creation (EWEA and Greenpeace, 2002: 20). The firms involved were mainly Spanish-owned. In 2002, the main centres of wind sector employment were Galicia (26 per cent), Navarra (16 per cent), Castilla-La Mancha (15 per cent), Castilla-Léon (14.3 per cent) and Aragón (13.7 per cent) (Río and Gual, 2006). These were also the regions having the largest concentrations of wind power capacity, the causal link being regional government policies.

In several countries, wind farms provide revenue for local authorities. In the Spanish case, Dinica (2002: 226) discussed the use of 'discretionary administrative royalties', whereby wind farm developers agree during the approval process to pay either a one-off fee or a percentage of lifetime profits to the local authorities. These receipts are recycled for community projects or for industrial investment. Local taxation revenues are also important in Germany and France, especially in economically depressed municipalities.

In all national cases, land rents for installation of wind turbines provide an economic inducement. They can be a significant source of revenue for some German farmers (EWEA and Greenpeace, 2002: 15). In struggling rural economies, such as Eastern Germany and parts of Spain, turbine rents have been an important source of income diversification for small farmers.

Local community benefits can also be engineered by a redistribution of revenue. In northern Germany, 'environmental compensation payments' divert a proportion of wind farm income to local nature protection funds. In Scotland and Wales, 'community trust funds' have been set up with energy companies to spread the benefits of renewables in the local area. However, this practice has aroused concerns, especially if agreement is related to a planning consent. Mechanisms for establishing 'trust funds', 'discretionary administrative royalties' and the like need not only to provide meaningful revenue, but also be based on transparent, accountable and impartial procedures.

Summary

The three main causes of social acceptance are present to a greater degree in the 'pioneer' countries than in the 'latecomer' countries. Further, the capacity and willingness of the latter to emulate the former is uneven. As liberal democracies, all five countries provide for community participation in planning processes, but specific responses to wind power have varied. As demonstrated in the cases of Denmark, Germany and Spain, spatial planning is well-suited to technologies involving extensive deployment, because it provides clear designations of suitability by zone. Further, the designation process gives meaning to public participation, since it leads to an identifiable, utilitarian output, namely a zoning map. Recognition of these advantages lies behind the 2005–6 French planning reforms which call for establishment of wind power zones through a 'bottom-up' participation process. The UK has been alert to the issues and the ODPM (2004b: 63) identified a 'new requirement for community involvement in significant planning applications'. However, it is still unclear how participation processes are to be improved and what identifiable new outputs can be expected. Indeed, tensions exist in the UK between statements extolling local participation and trends to centralised decision-making. The emphasis on regional targets led Toke and Strachan (2006: 161) to observe that 'the government is, in effect, seeking to gain adherence to its policy by issuing instructions from the centre rather than encouraging activism in support of wind power at the local level'.

Turning to community ownership, instances of local investment are rare in France and the UK, with wind farms generally developed by or sold on to large firms. As regards wider economic advantages, regional job creation in both countries is very limited since wind power technology is mainly imported. In the highly centralised UK fiscal system, there are few local tax gains. Land rents are paid of course, but often to large landowners. Thus whereas a number of local socio-economic benefits accrue in the 'pioneer' countries, few can be identified in France and fewer still in the UK. In consequence, it is extraordinary for British local authorities to *solicit* wind power development, whereas it is commonplace in Spain. In Denmark and Germany, close links were established between reaping local benefits from wind power whilst enduring its downside. Social acceptance was reinforced by perceptions of equity. On the other hand, in the UK and France the creaming of profits by distant third parties can provoke a sense of injustice and lead to social rejection. Even in Denmark, opposition arose to utility-owned wind farms. Similar hostility was predictable in 'latecomer' countries. In brief, the factors which explain social acceptance also help to understand rejection in those contexts where they are absent.

Conclusions

The characteristics of wind power technology, especially the impact of wind speeds, make judicious siting paramount and lead to regional concentration of build. But whilst wind is an inexhaustible resource, land is not. Cumulative effect and progressive onshore saturation are problems that require careful attention, given the high population densities that characterise much of Europe. The physical and geographical entailments of wind power challenge the institutional processes of planning, whilst creating social opportunities and costs. Acceptability criteria have been developed to respond to these factors. But because of geographical concentration, the challenges are spread unevenly, with a minority of local administrations and communities affected – and sometimes overwhelmed – by wind farm applications. Moreover, the 'fit' between national planning traditions and specific land needs for wind power deployment has varied. Where, as in northern Europe, zoning measures were already a part of the planning 'toolbox', wind farm construction was guided to low impact sites, with an increase in predictability for applicants, a reduction in tensions for (many) local communities and probably a decrease in transaction costs. On the other hand, a discretionary, criteria-based planning system – such as that which previously operated in France and continues in the UK – may simultaneously increase the apparent availability of land whilst decreasing the predictability of outcomes, giving a double incentive to multiply speculative applications. Although many proposals come to nothing, the process increases transaction costs and heightens community anxieties, provoking fears that 'nothing is sacred'. A further complicating factor in all reference countries is that whereas delays and failed proposals are often attributed to shortcomings in the planning regime, the true cause of outcomes often lie elsewhere, notably the lack of grid connections detailed in Chapter 6.

Embedded within planning systems as institutional processes for conflict resolution are opportunities to distribute costs and benefits in an equitable fashion. Three dimensions of the distributional equation were explored, namely societal participation in strategic locational guidance, community ownership and wider socio-economic advantages. These factors led to particular 'social contracts' in relation to wind power. They induced positive feed-back processes in Denmark, Germany and Spain, encouraging France to enhance at least the first dimension. The UK, however, scores low on all three. It is characterised by exceptionally limited community ownership and socio-economic advantages, yet has moved towards a 'top-down' decision-making process. Overall then, the 'pioneer' countries have sought and, to an extent, found the means to lower costs, increase benefits and distribute each with a measure of equity. Interaction between these factors enhances the social acceptance of wind power. This rounds out the explanation of why Denmark, Germany and Spain were successful in pioneering wind power, above and beyond their choice of policy instrument. These findings also point up the problems faced by 'latecomers', such as France and the UK, who need to undo the counterproductive institutional processes which store up social conflict.

8
Contesting Wind Power

Introduction

In discussing wind power, Gipe (1995: 322) observed that 'the pace of development alone can generate as much opposition as the manner of development'. Both the manner and pace of wind power deployment have differed in the reference countries studied here, but each has indeed aroused opposition. In Denmark, Germany and Spain, local communities were first acquainted with a fledging industry. Installations were small and ownership was generally local. As installations grew larger, communities learnt how to accommodate them progressively. Contrasts also emerged. In Denmark and Germany, local ownership was accompanied by a measure of green idealism, whilst in Spain utility ownership was tempered by local economic benefits orchestrated by regional authorities. Gradual development did not altogether avoid or eliminate conflicts, but gave time to manage them. Thus wind power deployment in the pioneer countries had in common a context of evolution.

In the UK and France, however, deployment occurred in a context closer to revolution. In both countries, calls to tender during the early 1990s led to the building of the first wind farms. But installations were few, small and scattered. With rare exceptions, they were constructed by outside developers to the curiosity, concern and sometimes hostility of the local population. Beyond these isolated experiments, wind power barely figured in the public consciousness. All this changed in the early 2000s due to new subsidies engineered by central governments to ensure rapid expansion of renewables. Meanwhile wind power technology had radically transformed. Turbines were larger, with first 1 MW then 1.5 MW models becoming the norm; towers were taller, with a height of 100 metres plus, and blades grew longer at 60 metres (or more). Installations became bigger, with the record for the largest wind farm being regularly broken. Ownership was by distant investors, often from abroad. In some communities, the combination of giant turbines in large farms developed by major consortia was experienced as intimidating and over-powering.[1] For opponents of wind power the sense of being crushed goes beyond 'visual impact' into the economic and political domains. Contrast this with Denmark, which has few wind farms and few giant turbines.

Consequently, the strongest hostility has arisen in the UK and France, but with contestation in the other three countries being visible in particular localities, circumstances and forms. The analysis that follows will therefore focus mainly on reactions in the latecomer countries, but with reference to parallel developments elsewhere. The aim is not to provide an exhaustive survey, but to identify representative elements of contestation and to explain their motivations, arguments, objectives and strategies. The chapter's first section deals with antiwind groups whose reason for existence is to oppose wind power. The second section analyses the reactions of long-established organisations who, for contingent reasons, have objected to wind power proposals but hold no brief to stop its spread. Thus the term 'contestation' is used to refer to two distinct categories of actors. By comparing and contrasting the two groups, the third section will demonstrate that they do not share a common agenda or goals. Nevertheless, in combination the forces of contestation have opened up the wind power, energy and climate debates and alerted decision-makers to emergent policy needs.

The arguments and activities of antiwind protest

Local resistance to the erection of wind turbines first appeared in Denmark during the 1970s (Farstad and Ward, 1984: 91). However, opposition was mostly directed against large, utility-owned projects (Nielsen, 2002: 128). Danielsen (1995: 61) noted that by the 1990s 'groups of citizens living close to planned wind farms have protested that they destroy the landscape and the noise from so many turbines makes wind farms highly undesirable neighbours'. To a significant extent, both the construction and the contestation of wind power were matters for local communities. Indeed, the Danish umbrella association formed to resist inappropriate turbine construction and operation is named 'Neighbours against Windmills'. It has worked to alleviate problems for residents (such as noise, disturbance and so forth), but was also critical of government policy as over-generous. The reforms of the 2000s, which slashed subsidies and precipitated collapse of onshore build, appear to have diminished its grounds for contestation. In Germany too impacts on residents have formed the typical grievance for antiwind groups. Activities include mobilisation through meetings and petitions, objections to consents and legal challenges to suspected illegalities. These groups are typically local in scope, but national coordination is gradually emerging, for example through the *Bundesverband Landschaftschutz* (BLS). In Spain, local pockets of resistance have arisen for similar reasons, whilst the harm caused to bird populations in Tarifa and Navarra has become a specific rallying point.[2]

The UK has seen the development of a larger current of anti-wind protest, having broad ambitions. Three 'umbrella' organisations operate at the national level. The oldest is Country Guardian which, since its establishment in 1991, has systematically updated its case against wind farms (Country Guardian, 2000b; Country Guardian/Etherington, 2006). Expressing concerns over the environmental and social impacts of wind power, it provides information to protest groups in the UK (Country Guardian, 2006), including advice on 'how to fight a wind-farm' (Country Guardian, 2004). With rapid expansion of wind power in the 2000s, two further umbrella organisations were formed. Views of Scotland (VoS) believes that wind power is causing 'unjustifiable and irreversible damage to some of Scotland's greatest assets' (VoS, 2006). It campaigns for 'a secure, sustainable energy policy that respects the rights and aspiration of all citizens' (VoS, 2004: 1). Its activities include news gathering, production of reports, giving evidence to official inquiries, lobbying politicians and providing support to local groups. In England, the Renewable Energy Foundation (REF) claimed that wind power deployment amounts to 'an industrialisation of the countryside and the destruction of our most precious heritage' (REF, 2004a). It aims to raise public awareness through research, lobbying and legal/adminis-trative support to local groups, and to change policy on renewable energy, arguing that a bias towards wind is detrimental to other renew-ables (REF, 2004a).

In France, fears of a 'wind rush' led to the establishment of *Vent de colère* (VdC) in 2001. Many of its criticisms of wind power are compara-ble to its UK counterparts. But its stance on energy supply and climate change issues is distinctive. Noting that 90 per cent of French electric-ity is already generated from GHG-free sources, it argues that wind power brings no climate change bonus, since replacing nuclear or hydro with wind is GHG neutral. Its leaders argue that wind cannot provide an exit from nuclear power, due to the intermittency of the resource and that wind power is surplus to requirements since French electricity generation exceeds domestic consumption (VdC, 2004). The organisation is critical that climate policy has focused on electricity generation rather than transport, industry and housing, which are larger GHG sources in France. Thus it redirects – rather than dismisses – climate change discourse, accusing government of policy errors. It is unsparing in its critique of the feed-in tariff, claiming that it generates undeserved profits for wind developers at the consumers' expense, whilst harming local communities (VdC, 2004).

The main arguments of antiwind protest

Broadly speaking, criticisms of antiwind groups fall under three main headings: technology choice and energy policy, environmental issues and amenity issues. These will next be summarised.

Technology choice and renewables policy

Opponents have often claimed that wind technology is a poor choice because it fails to deliver. This view was already voiced during the 1980s about wind farms in Altamont, California. Residents saw blades which stood idle, and this lack of physical motion was taken as proof of wasted investment (Thayer and Freeman, 1987: 394). The theme of a failing technology is expressed through a number of variations: generation is marginal, load factors are low, production is impaired by unpredictability and lack of 'firm capacity'. According to Country Guardian (2003) 'the fatal flaw is intermittency', leading to diminished security of supply and increased costs (and emissions) due to fossil fuel back-up. Views of Scotland claimed that wind power 'suffers from low energy density, inability to store energy, random intermittency and finite probability of common-mode failure' (VoS, 2004: 3). Associated problems cited by these organisations are the risks of interruption of supply, grid instability and transmission losses in moving electricity from rural sources to distant load centres. UK antiwind groups often seek to support these contentions by reference to critical reports produced by continental grid operators and utilities. A frequently cited source is the German TSO, E.ON Netz (2004, 2005).

Charges of ineffectiveness slide over into assertions that wind power represents a ruinously expensive and misdirected investment. Country

Guardian (2003) has majored on wind power as a 'white elephant' – namely a waste of resources. The REF (2004b: 10) argued that the RO created a 'grave distortion' by directing investment to wind power which was 'of least value and use', rather than to reliable, dispatchable sources. Drawing on critical analyses by CRE (the French electricity regulator), VdC (2006: 4) condemned subsidies which, it considered, led to huge profits for developers.

The scale of recourse to imported technology has also prompted criticisms. Opponents have attracted attention to the very different industrial and employment conditions characterising wind power in the UK and France, as compared to the pioneer nations which manufacture and export the technology. In relation to the Causeymire wind farm in Caithness, Views of Scotland (2004b: 9) reported that 'staff boasted that every last nut and bolt was Danish-made, that the site was erected by Danish engineers and that it is even controlled from Denmark'. *Vent de Colère* (2006: 6) stressed the balance of trade deficit caused by imports of wind turbines: target capacities for 2010 of 10,000 MW (or more) would cost in excess of 10 billion euros, yet create few jobs in France. Critics also contend that wind farms have adverse impacts on economic activities conducted in the vicinity of wind farms, particular tourism onshore and fishing offshore. Thus the broader critique has been that policy lacked coherence, amounting to a 'short-term vision of unsustainable development, the consequence of a walk in the dark' according to Views of Scotland (2004: 4).

Environmental issues

Conservation issues are prominent in antiwind discourse. The leitmotif of Country Guardian is protection of the countryside from industrialisation by wind farms. In its manifesto, Country Guardian (2000a) aimed to legitimatise its stance by reference to the 1968 Countryside Act and its stress on 'conserving the natural beauty and amenity of the countryside'. Since the peat slide at Derrybrien (Ireland),[3] the protest organisations have been highly critical of wind farm construction in peat bogs (which are a form of 'carbon sinks').[4] When challenged on the implications of climate change, the position of antiwind groups appears to have evolved. Rather than deny or minimise the existence of climate change, the current strategy is to cast doubt on the capacity of wind power to deliver significant emission reductions.[5] This strategy has also been used by antinuclear groups to support their rejection of nuclear power.[6]

Amenity issues

Amenity issues represent the third main plank of antiwind protest. They arise in relation to undisturbed enjoyment of the countryside, but considerable stress is also placed on the living standards of home-owners close to wind farms. Concerns relate to nuisance and distress caused by visual impact and noise, safety and depressed house values. A theme acquiring greater salience is discord created within rural communities by developer proposals, since rent-seeking landowners reap the benefits whilst neighbours endure the costs.

The action repertoire of antiwind protest

Antiwind protest groups have sought revisions in planning guidelines which, they believe, have been rendered obsolete by changes in wind power technology. They seek greater protection for nature and for human beings by exclusion of wind farms from valued landscapes and wildlife habitats, specified separation distances from housing and revised levels of permissible noise. (This mirrors and responds to the strategy of the wind lobby which has sought to instigate a planning regime that is more favourable and responsive to their own preferences.) One bone of contention for UK protestors is that developers can appeal against rejection of a planning application, but the public cannot appeal where consent is granted.[7] Critics have claimed that this, together with increasing recourse to section 36 of the 1989 Electricity Act, marginalizes the elected local authorities and the public, creating a 'democratic deficit'.[8]

A major tactic of protest groups is to object to planning applications, with the aim of preventing the construction of a wind farm. As noted by Toke (2005b: 1535), the main movers behind local campaigns are usually those living closest to the proposed site, with the consequence that the attitudes of immediate residents and their parish councillors 'have a major influence on the planning decision outcome at the local authority level'. During the planning procedure, protestors will typically contest the accuracy of the EIA report as prepared by the developer (see p. 144). Because the EIA report covers a range of separate material considerations, antiwind groups tend to respond under each heading in order to maximise their case. This involves a division of labour whereby group members specialise in contesting findings in one area, be it noise and nuisance, landscape impacts, effects on wind life and so forth. Their strategy is rounded out by an information campaign to mobilise local people and motivate them to participate, sign petitions and send in letters of objection. Although opinion polls on wind power typically show large majorities broadly in favour, the reservoir of potential discontent available for tapping at the local level is frequently misunderstood and underestimated by the prowind lobby. In many cases, members of the public who initially favoured wind power become disaffected once a proposal materialises on their doorstep and they learn first hand about the technology, its deployment, the tactics of developers and the consequences for their locality. The antiwind repertoire has particular entailments: because local campaigns are focused on planning hearings, their lifetime is limited to opposition to a single project and the discourses of protest tend to be fragmented and fault-finding. This leaves antiwind groups open to charges of a negative agenda, raising the question of their overall goals.

The agenda of antiwind protest

The contentions of antiwind groups are often expressed in uncompromising terms, contributing to conflictual relations with developers. This encourages a heated exchange of claims and counter-claims between pro and antiwind groups, regarding the activities and aims of each. Identifying the goals of antiwind protest is harder than defining those of the prowind lobby. The latter takes up a maximalist position, in other words seeking to construct maximum capacity in the shortest time. But does antiwind protest adopt the minimalist position of seeking to stop all build?

The aim of local antiwind groups is generally to prevent the construction of a specific wind farm. This, however, does not warrant the inference that all opponents would prefer no construction anywhere, since objections are lodged on the basis of local conditions – excessive proximity to housing, encroachment on conservation areas or common land, harms to rare birds and their habitats, and so forth. An approach based on *conditionality* is, of course, at the heart of the UK planning system. Consents are given where criteria are met and refused when not. Having expressed reservations based on local conditions, objectors are under no requirement to specify desirable locations for categories of installations – be it wind farms, roads, or any other construction. Indeed, to the extent that such statements might prove counter-productive during local deliberations, objectors may have an incentive to avoid them altogether. Thus it is inadvisable to 'read off' a general agenda from the repertoire of local antiwind protest, or ascribe a single aim to *all* individuals involved.

Yet systematic opposition can characterise a minority of protestors – especially within the national 'umbrella' organisations – although the forms it takes need qualification. In its manifesto, Country Guardian (2000a) stated that it is not opposed to 'wind energy per se' but to 'commercial wind power', and specified conditions of acceptability for usage of the technology related to environmental conservation, safety, nuisance, community harmony, property values and economic livelihood. However, the insistence that wind installations be 'sited away from the grid' (Country Guardian, 2000a) precludes at face value all large-scale deployment (including offshore). This choice of wording suggests a generic 'no to wind'. Even more hard-line is the position of *Vent de Colère* (2005) which rejects 'all forms of industrial-scale wind power' ('*la position immuable de VdC est de refuser toute forme d'éolien industriel'*). Other 'umbrella' organisations object to the pace and scale of commercial wind deployment, but seem to have avoided outright public condemnation of grid-connected installations.[9]

Whilst protestors cannot all be pigeon-holed as sharing one, identical outlook, the interaction between the prowind lobby and antiwind groups has clearly led to polarisation. At one extreme are partisans whose 'yes to wind' is a categoric affirmation. Amongst their number are individuals who also categorically state a 'no to nuclear'. In the antiwind camp, a small number of protestors can be identified whose 'no to wind' is categoric. But does this 'mirror reflection' extend to support for nuclear power on the part of antiwind protesters? Antinuclear partisans have repeatedly made the charge that antiwind groups are funded by the nuclear lobby, but have failed to provide proof.[10] Polarisation is also evidenced within the truth claims of both pro and anti factions. Antiwind protestors use colourful language to describe their adversaries, including charges of 'green window dressing' (Country Guardian, 2000b), 'intentional deception' (Halkema, 2006: 21), 'scam' (VoS, 2003b) and '*arnaque*' meaning confidence trick (VdC, 2006). In response, the prowind lobby has recourse to the rhetoric of 'myths'.[11] Each side maintains that the claims of the other are misleading, a half-truth or plain fiction.

Why is the trend to polarisation so pervasive in the wind power debate? One explanation arises from the phenomenon of 'devil shift', by which Sabatier and Jenkins-Smith (1999: 140) understood the tendency of actors in high-conflict situations to perceive their opponents as more powerful and/or more evil than they are. If the opponents are evil, then their victory is likely to result in very substantial costs. And if the opponents are powerful, the only way to preclude their victory is to achieve highly effective coordination among the like-minded. Thus 'devil shift' expands from perceptions and attitudes into a strategy for collective mobilisation. But whilst catalysing joint action, 'devil shift' propagates deformed stereotypes. Examples are the 'nuclear bogeyman' and the 'fossil fuel dinosaur'. These caricatures of technology choice have been imported wholesale into the wind power debate, notably by the NGO wing of the prowind lobby. Antiwind protest has responded in kind, by portraying wind turbines as ravenous industrial giants despoiling the countryside. Thus the wind power debate illustrates the observation that 'discussions of our energy options too often simplify the world into good guys and bad guys' (Jaccard, 2005: 254), producing a climate inimical to reasoned inquiry whilst favouring ill-considered decisions. Further, the often angry and intemperate language of anti-wind protest may have alienated long-standing organisations which have their own reservations regarding wind power deployment.

The objections of established organisations

Many established bodies have made criticisms of wind power. However, the extent of their opposition has varied along dimensions such as the timing of their intervention, the propensity to public disagreement, the content of their objections and their criteria of acceptability. Analysis of the range of objections and their impact on wind power deployment rates is complicated by the development process itself. Developers initiate proposals, set up anemometery and undertake 'scoping' before making a planning application, which is then deliberated. On a schematic (and necessarily simplified) basis, the development process can be divided into four stages: pre-application, application, approval and installation. In practice, a proposal may fall at any of these stages, for reasons that never fully emerge into the public domain. A common – but singularly incomplete – view is to blame 'slow' deployment rates either on antiwind protest or on the planning system. However, a large number of proposals never proceed to a formal application, for reasons such as disappointing wind levels or logistic problems.[12] But factors such as lack of grid connection, investment shortfalls or the identification of better prospects elsewhere can lead to cancellation of proposals at all stages up to and even beyond the point where a consent is given. Hasty conclusions on deployment rates can be avoided by probing the range of influences affecting the translation of wind farm proposals into working installations.

In the UK, it is considered good practice for developers to liaise with 'statutory consultees' prior to a formal application. The 'statutory consultees' cover a range of organisations, among which government agencies figure prominently. They include the Ministry of Defence, Scottish Natural Heritage and Natural England.[13] When consulted, these agencies will give a view on the acceptability of the proposal. Because this process is guarded by 'commercial confidentiality', the public rarely discovers what proposals are made, how the consultees respond, or what motivates the developers to accept or reject the feedback. This process complicates analysis of why wind farm proposals stand or fall, and whose influence proves determinative.

The problem of electromagnetic interference, particularly upon radar equipment, is the prime example of a low-profile but crucial determinant. Simpson (2004: 49) reported that the UK Ministry of Defence opposed almost half of all proposals for wind farms submitted in 2003. The BWEA acknowledged that around half of potential sites were affected by concerns over interference with military or civil radar (Massy, 2006: 25). France also has a large number of wind farm proposals affected by radar exclusion zones (Dodd, 2006), whilst in the USA around 1000 MW of development were put on hold in 2006 pending resolution of radar issues (Anderson, 2006). The scale of the problem has led military and civil authorities to seek 'technical fixes' for the future. But in the recent period, perhaps the main 'roadblock' impeding wind power proposals – often upstream of a planning application – has been radar interference.

Nature protection agencies and organisations have also proved instrumental in channelling wind power development. Ornithological associations have identified a range of impacts of wind turbines on birds (and bats), notably collision mortality, loss or damage to habitats, disturbance (including barrier effects, displacement and deviations from migration routes) and effects on reproduction rates. These factors lead to adverse and additive impacts on populations. It must be stressed that impacts are non-generalisable – being species and site specific – but have proved significant in particular locations, notably Altamont, California and Tarifa and Navarra in Spain.[14] In relation to the unfolding impacts of wind power, the ornithological societies are challenged by the need to improve their methodologies and data bases across a range of contexts and with limited resources. Striking a balance between short and long-term threats to wild life on the basis of bounded knowledge has placed the ornithological NGOs in a dilemma. The Royal Society for the Protection of Birds (RSPB) takes the view that climate change is 'the most serious long-term threat to wildlife in the UK and globally', and 'supports the increased use of wind power, as long as wind farms are sited, designed and managed so they do not harm birds or their habitats' (RSPB, 2004). In association with other groups, it produced guidance on how this can be achieved (English Nature, RSPB, WWF-UK and BWEA, 2001). Notwithstanding its support in principle for wind power, the RSPB has objected to around 10 per cent of UK onshore sites, according to Toke (2005a: 51). On the announcement by the UK government of a second round of offshore wind farms, the RSPB (2003) voiced anxieties over overlap and proximity to major bird habitats. The organisation seeks resolution of its concerns on the basis of scientific research leading to appropriate mitigation. However, high-profile instances have arisen where mitigation may not prove possible. The RSPB (2005) objected in strong terms to a project to install 702 MW of wind power on the Isle of Lewis, due to its extensive impact on bird habitats protected under European designations. Likewise, in Spain SEO/Birdlife (2005) declared itself in favour of wind energy, yet objects to its consequences for protected habitats in Valencia, Extramadura and elsewhere (SEO/Birdlife, 2006a, 2006b).

Given its remit to ensure sustainability, Scottish Natural Heritage (SNH) has sought to limit environmental damage caused by climate change. Although in principle favouring wind power, it has recognised the need for the climate programme to extend well beyond renewables deployment (SNH, 2002). The agency has stated that it is 'concerned at the number and scale of onshore wind proposals which have either been submitted for consent, or are at a preparatory planning stage' (SNH, 2005b), a concern arising from the potential to cause extensive change in rural landscapes with impacts on significant species and habitats. To help resolve the tension between protection of natural heritage and its acceptance of renewables, SNH (2005a) prepared 'strategic locational guidance' for wind farms in Scotland, categorising zones by level of natural heritage sensitivity and seeking to guide deployment to least sensitive areas. It is also developing a methodology to avoid 'cumulative effect' within the latter areas, since they lie closest to centres of population (SNH, 2003). This balancing act between conflicting desiderata has led the agency to object to a significant number of proposals. SNH (2004: 19) reported that in relation to 67 onshore applications between 2001–4, it accepted 40 per cent without objection, formally objected to 21 per cent and made a 'conditioned' objection (placing requirements for improvements on the developer) to 37 per cent.

In England, the major countryside agency has repeatedly voiced reservations over wind power deployment. Previously known as the Countryside Commission (1991: 15 and 9), it declared that there was 'a fundamental incompatibility between such developments [e.g. renewable energy schemes] and the protection of the countryside', calling for 'the most stringent environmental standards'. In relation to wind farms, it proposed a policy based on what was 'environmentally acceptable' which, in its view, excluded construction in national parks, AONBs and heritage coasts, and recommended a separation distance of 300m in relation to housing (Countryside Commission, 1991: 2 and 11). It stressed that 'all too often the treatment of landscape impact is seen as a purely "subjective" matter rather than one of informed judgement based on a combination of both objective and subjective methods' (Countryside Commission, 1991: 15). To deepen this approach, its successor the Countryside Agency – now subsumed under Natural England – developed a program of landscape classification and evaluation. In recent years, the Countryside Agency intervened sparingly in the wind debate but objected to problematic applications, with its intervention in the public inquiry on the Whinash (Lake District) proposal being a prominent example.[15]

Landscape protection NGOs have been instrumental in resisting the spread of wind power. In the UK, the Council for the Protection of Rural England (CPRE) and the Council for the Protection of Rural Wales (CPRW) are long-standing associations which have been highly critical of the effects of wind farms on the countryside and on rural populations. In France the Société pour la protection des paysages et l'esthétique de la France (SPPEF) has adopted similar positions. At the European level, Europa Nostra – which brings together national NGOs sharing the aim of preserving natural and cultural heritage – has expressed alarm. The remit of these organisations has forced them into the wind power debate. Seeking to remain true to their traditional values regarding the preservation of rural locations and prized landscapes from urbanisation and industrial development, they have also been forced to acknowledge new threats to the countryside posed by climate change. Whilst varying on the specifics, their responses have been based on conditionality and provide commentaries on the circumstances under which wind power development is acceptable or not. Some of their commentaries are extensive, and only key points can be summarised here.

Europa Nostra (2004a) denounced 'serious damage to the environment' caused in countries which 'have provided heavy incentives for development of wind power, relaxed planning legislation and failed to make a balanced assessment of its merits', and offered a list of considerations on how 'balanced assessment' can be achieved (Europa Nostra, declaration, 2004b). Whilst stressing its opposition to 'anarchic' installation of wind turbines, the SPPEF (2002) specified acceptability criteria in terms of location (*inter alia*, exclusion from national parks, designated sites and 'emblematic' landscapes), and also in terms of administrative procedures (SPPEF, 2006). The CPRE (2006) acknowledged the danger of climate change, but saw reductions in energy consumption as the principal means to effect GHG reductions, stressing that 'while CPRE will support renewable energy development in certain cases, such schemes should not come at the expense of the countryside (...) we will strongly resist those which damage the beauty, tranquillity and diversity of the English countryside'. Its acceptability criteria emphasise 'a sequential approach (...) to steer wind development to the least environmentally sensitive areas and encourage development on brown field sites', as well as calling for greater community participation in decision-making (CPRE, 2006). The CPRE is prone to object to wind farm planning applications, with considerable success according to Toke (2005b: 1531–2) who found that in his sample of planning outcomes 'there is not a single case where the Planning Authority has approved permission after an objection by the relevant landscape protection organisation, usually the CPRE'. The CPRW (2005a) lamented the creation of 'a distorted and divisive conflict between the two aims of reducing harmful emissions and protecting the high quality rural landscape of Wales', whilst sharply criticising renewables policy in that it 'fails the fundamental premises which underpin the concept of a sustainable and environmentally responsible energy portfolio' (CPRW, 2005b: 2). Considering itself to be a longstanding practitioner of sustainable development, CPRW rejected what it considered to be a inappropriate exploitation of the concept. Because current policy – in its view – gives unjustified dominance to a single technology (wind power), the CPRW (2005b) offered a detailed alternative scenario for energy sourcing and consumption.

Certain UK amenity organisations have taken a positioning similar to the landscape protection NGOs. As part of its long-standing remit to protect common land, the Open Spaces Society (OSS) opposes the siting of wind farms on common land; for other locations, it proposes 'tests' of acceptability related to designations, landscape characteristics and public rights of access (OSS, 2006). The Ramblers' Association (RA) has opposed the 'damaging of upland landscapes with heavy

engineering projects like wind turbines' (RA, 2006), and ran advertise-
ments in national newspapers during the 2005 general election to draw
public attention to the dangers. In calling for recognition of the 'value
of the countryside for its own sake' (RA, 2005), it asked for environmental safeguards in relation to designated sites and high quality landscapes, and offered alternative energy sourcing scenarios.

In summary, major and highly respected organisations have objected
to wind power. However, their reasons for objection have been contingent on their remits, on the locations proposed and on the availability
of mitigation. Thus far from offering systematic opposition, they have
evolved a complex range of acceptability criteria to avoid, resolve or
mitigate the problems encountered.

The absence of a common front

The existence of a variety of organisations which repeatedly raise
objections to wind power may suggest a common front. But the reality
is quite the opposite. In practice, joint and concerted action against
wind power rarely arises, even where organisations lodge objections to
the same planning application. This is because the actors involved are
not motivated by a common cause, whilst overlap between their aims
tends to be coincidental rather than strategic. To understand why this
is so requires exploration. Two core factors can be identified: the lack
of common 'story lines' and a divergence of agendas.

As discussed in Chapter 3, 'story lines' are 'the medium through which
actors try to impose their view of reality on others, suggest certain social
positions and practices, and criticize alternative social arrangements'
(Hajer, 2005: 304). On the basis of common 'story lines' actors will form
'discourse coalitions', which may or may not lead to concerted action in
the field. With regard to wind power, multiple strands of contestation
have led to a plurality of 'story lines'. Drawing on the preceding empirical discussion, the following 'story lines' are observably the main candi-
dates for the promotion of convergence among their bearers:

- 'intermittency and unreliability of the technology';
- 'policy imbalance';
- 'the democratic deficit';
- 'the industrialisation of the countryside'.

However, close investigation shows that the federating effects of these
'story lines' are often greater in appearance than reality.

The short-comings of the technology have exercised antiwind groups, as well as utilities, transmission operators and regulators in some (but not all) of the reference countries. Indeed, UK antiwind protest systematically draws on the commentaries, criticisms and (historical) hostility of the German utilities to wind power in order to provide expert support for its case on 'intermittency and unreliability'. However, underlying motivations are divergent. The position of German utilities is not consistently 'antiwind', but contingent upon prevailing policy conditions. Hoppe-Kilpper and Steinhäuser (2002) hit the mark in observing:

> The electricity utilities complain that under the electricity feed-in law they face an unfair burden of paying artificially high prices. Theirs is not a fundamental rejection of wind power but rather a desire for a guaranteed method of compensation.

This observation is supported by the UK case-study, where the generous payments offered by the RO have incentivised the German utilities to take a major role in wind farm construction. A more rounded characterisation of the position of electricity industry actors is that 'intermittency' is an operational problem that is manageable given enough technical and economic resources. Thus they do not support the 'story line' that 'intermittency' is an intractable flaw of wind power – at least not at current and near-term penetration levels. Further, practitioners who are inconvenienced by the shortcomings of the technology tend to seek 'technical fixes'. This is true not only with regard to generation and transmission, but also in relation to the concerns of civil and military authorities over electromagnetic interference with radar, television broadcasts and so forth.

The 'story line' of a 'policy imbalance' – namely, the monopolisation of support to renewables by wind – has met with echoes in many quarters. But once again this is a highly contingent claim. It has little foundation in some national contexts (for example Germany, which has offered hefty support to renewables such as PV). Although the claim is better grounded in the UK, a broadening of the bases of policy support was, at the time of writing in early 2007, a likely consequence of the 'energy review'. Such correction is a normal part of the policy process, especially in the early stages of a learning curve in technology development. Calls for adjusting the policy balance will be persuasive at particular junctures, but (except with the worst policy blockages) will not mobilise on a long-term basis.

A 'democratic deficit' has grieved a number of parties critical of the planning system and of the scope it offers for local participation – in relation not just to wind farms but also other developments. The mobilisation potential of this 'story line' may yet prove considerable on a cross-sectoral basis, namely above and beyond wind power *per se*. However, it has not provided a federating agenda in relation to wind because a major group of UK objectors have privileged access to the system. These are the 'statutory consultees' who, by definition, must be consulted. In a number of cases, they are mandated government agencies. These organisations do not suffer from a participatory deficit. They can exercise upstream influence on proposals to the extent of scotching some entirely, with the Ministry of Defence being the main case in point in the UK (with comparable outcomes in other countries). Where planning applications are brought forward, their views are usually respected by local authorities. Thus protestors may seek to catalyse these organisations into registering an objection to a specific proposal, but have little scope to recruit them into a common front to redress a 'democratic deficit'.

On the other hand, the 'story line' of an 'industrialisation of the countryside' has acquired salience and federating force. So long as wind power capacity was counted in the hundreds of megawatts, this 'story line' could be dismissed as exaggeration by wind power enthusiasts and as a bizarre irrelevance by the majority of the population who had never seen a wind farm. But now that turbines number many thousands in several European countries and first-hand experience of their extensive spread is gained by a growing cross-section of the public, the power of the 'story line' has augmented. This discourse draws part of its power from a tradition of nature conservation and anti-industrialisation going back to the nineteenth century.[16] The 'story line' is sharpened by the increase in environmental pressures experienced in recent decades due to urban sprawl, industrial pollution and intensive agriculture. These developments add substance to a core contention of antiwind protest which is that the social and environmental costs of wind power are real, and cannot be discounted to zero as enthusiasts sometimes claim. The 'story line' is also abetted in the UK by growing disaffection in rural populations at the perceived indifference or incompetence of central government in handling 'countryside' matters such as 'foot and mouth' disease, development control and so forth. In consequence, this 'story line' is central to the discourse of the antiwind organisations, the landscape protection and the amenity associations. It finds echoes in the conservation agencies and in segments of the tourism sector dependent on the draw of scenic landscapes. It also has resonance for the ornithological organisations, since the habitats of protected species are under increased threat. These factors have produced limited signs of joint action, for example between antiwind protest and the CPRE (and its cognates in Wales and Scotland).

However the industrialisation 'story line' is not accepted by all parts of the environmental movement. International NGOs such as Greenpeace and Friends of Earth have not only supported wind power deployment, but distanced themselves from countryside concerns. This change in stance is highlighted by Dunion (2003: 23), an FoE activist:

> In many respects, Friends of the Earth Scotland typifies an adjustment in the focus of the green movement (...) There is little in our work now which is the conservationist agenda – previous issues such as resisting the culling of seals or whales, damage to peat bogs or conifer afforestation have gone from our agenda and we are not engaged with scrutinising biodiversity programmes or site designations. It is not because these are unimportant. It is simply that ours is an environmentalist agenda, harnessing environmental health and social justice to the sustainability issues of living within our environmental space.

A three-way split within the 'green movement' between conservationism, environmentalism and ecologism is well-known amongst activists and extensively documented in the academic literature.[17] On the other hand, the general public still tends to view 'green' as one colour, rather than several. The split has consequences for the wind debate in that the international NGOs who focus on the abstract 'environmental space' of climate change (and so promote wind power) dismiss the localised conservation concerns of the landscape, habitat protection and amenity organisations (who often resist it). The ensuing stand-off serves both to deepen the split and draw attention to it. It induces a painful posture for organisations straddling both sides of the divide, notably the ornithological NGOs. Organisations such as the RSPB share a common front in terms of discourse and policy aims with Greenpeace, FoE, WWF *et al* in relation to climate protection. However, the RSPB, the SEO/Birdlife and their cognates remain wedded to a local/national conservationist agenda revolving around biodiversity, site designations and preservation of habitats – the very elements which have dropped out of FoE's agenda entirely or persist only in relation to distant continents (WWF) or ocean marine life (Greenpeace).

The re-opening of the rift in the environmental movement raises questions over the content of sustainable development. Twenty years ago, Michael Redclift (1987: 200) confidently asserted that:

> in seeking sustainability in the North we are seeking to affirm a cluster of related values, concerning the way in which we want the environment to be preserved. We seek, with millions of other people in the developed world, to protect and conserve rural space, to recognise aesthetic values in the countryside, to provide better access to this space and to ensure the biological survival of threatened species.

Here the inscription of the values of conservationism onto sustainable development foresees no tension between the two. Yet a disjunction is precisely what the climate coalition NGOs now stress. At the risk of simplifying their position, they appear to prioritise the safeguarding of the 'global commons' over and above local environmental protection: if necessary, the latter can be sacrificed to the former on utilitarian grounds. However, the necessity of any such sacrifice is rejected by critics of wind power. In their view, climate policy should not add yet more environmental costs, especially if it should prove ineffective. Disagreements over the aims of climate policy and the content of sustainable development are important components of the wind power debate.

This sharpens the question of the underlying agendas of the various organisations objecting to wind power. We have seen that systematic opposition to wind power is expressed by at least some antiwind protesters. Whilst systematic rejection is *not* a characteristic of all local and national groups, and some participants stress this explicitly,[18] structural and strategic determinants frequently tilt the scales in this direction. The adversarial nature of planning hearings and public inquiries pushes local groups towards an unreserved call for denial of consents. For national umbrella organisations (who are relatively new arrivals on the scene), the construction and communication of an identity and a programme are facilitated by a univocal stance. After all, their reason for existence is to oppose wind power. The consequence is that antiwind protest comes to acquire a negative agenda.

However, long-standing bodies who object to wind farm proposals on an *ad hoc* basis tend to distance themselves from antiwind groups. Their reasons are not broadcast publicly, and so can only be inferred. One is presumably to disassociate themselves from a negative agenda.

Another is that government-sponsored nature conservation agencies are enjoined with an arbitration role requiring objectivity and impartiality, and so are averse to systematic intervention on either side of a partisan debate. Other affected parties, such as ornithological associations and tourism boards, likewise need to maintain autonomy of action. Objections by these organisations are contingent, arising in many cases from impediments to their normal operations. Such reservations can often be overcome by 'technical fixes' or other types of mitigation which, once in place, allow objectors to resume their 'business as usual'. The existence of a highly specialised mission channels their responses. The ornithological associations do not have a calling to protect all birds, only threatened species. Their intervention is limited to cases of severe threat by wind farms to rare birds in protected habitats. Outwith such areas, their motivation for involvement drops sharply, due to the dilution of their limited resources in contexts of diminishing returns. The landscape protection NGOs seem to have drawn closest to antiwind protest, particularly in the UK. However, there is a structural reason for this. The overlap in the UK between the high wind zones sought by developers and the scenic upland landscapes deemed worthy of protection is very high. This contrasts with Denmark and northern Germany, where wind farms have often (though not exclusively) been sited on low-lying, agricultural land. Further, organisations such as CPRE have a century-long tradition of resistance to industrial incursions into the countryside: thus their objections to wind power are consistent. However, theirs is a positive agenda to preserve cherished landscapes for their intrinsic value to present and future generations, rather than a systematic rejection of wind power as a 'flawed technology'. In summary, the underlying agendas and motivation of objecting organisations diverge at varied but distinct junctures from the systematic opposition to wind power that characterises the 'hard core' of antiwind protest. This context has favoured the emergence of a loose constellation of actors who are far from presenting a common front.

Conclusions

A number of organisations contesting wind power have raised a cry of alarm over its effects on landscape, wild life and human beings. They have done so in an era of heightened anxiety over climate change and energy sourcing. Their intervention has raised a broad question: who is empowered to speak to this range of issues and in what fora? The emergence of new actors in the wind power debate has impacted on the 'traditional' answers to this question. Energy policy in the past was the restricted domain of an administrative, scientific and corporate elite, with little input from civil society. In many respects, policy communities relating to renewables still display this limited and closed membership. Environmental NGOs have, however, extended societal participation in climate and energy debates. Yet the historical opposition to nuclear power of Greenpeace and Friends of the Earth strongly colours their energy policy recommendations. Their campaign strategies deliberately encourage partisanship in relation to energy sourcing options. Whether such organisations hold a mandate from the public for their advocacy is an interesting question. The wind debate has, however, also pulled in other categories of NGO into energy and climate policy, notably the countryside protection and the amenity associations. In responding to new threats, these organisations have remained true to their historical mission. To avoid charges of being antiwind and promoting a negative agenda, some have proposed alternative energy sourcing and use strategies. This has moved them considerably beyond their long-standing remit. It has left them open to charges of overstepping their mission, of lacking the necessary expertise or of diverting resources from core activities.

New organisations with a vocation to carry energy policy debates forward have been slow to emerge. Exceptions include groups to support or reject wind power. However, the prowind groups, such as 'Yes2Wind' or 'Embrace the revolution' have so far served as a 'shop window' for lobbyists from the BWEA and from anti-nuclear NGOs, flying the colours of one component of the environmental movement. On the other hand, the leaders of organisations such as Country Guardian and *Vent de colère* seem focused on the negative agenda of stopping wind power in its tracks. Yet antiwind groups have forced a reaction from both institutional and established NGO actors. Though their dissident views, antiwind groups have enlarged the debate. In the process, the declarations of their umbrella organisations have modified. Whilst early statements from both sides sometimes contained identifiable errors, greater attention is now paid to avoiding statements that can be easily disproved. These developments demonstrate an on-going social learning process arising from discursive interaction between competing sources of advocacy, which in turn encourages more members of the public to engage with energy sourcing and climate challenges.

However, at their core, pro and anti factions have very restricted aims. There is considerable need for outward-looking, non-partisan analyses in order to stimulate debate, mobilise the public, contribute to policy-making and lead to improvements in energy sourcing and use. The debate on wind power cannot be limited to unconditional acceptance or rejection, but is about setting boundary conditions for acceptability of one technology among many.

9
Reviewing the Outcomes: Policy Learning and Path Choices

Introduction

This final chapter aims to synthesise earlier discussions and open out the debate. Its first section takes stock by summarising key findings related to the wind sector. The second section reviews progress in policy learning and makes policy recommendations. In the third section, the question of the development path taken by the wind sector is addressed, leading on to discussion of future options for renewables and changes in energy sourcing more generally. In particular, the scope for new 'social contracts' in energy sourcing is explored and the implications for sustainable development are considered.

Taking stock

Chapter 2 made a number of findings regarding industrial leadership, ownership models and development paths in the wind sector. The pioneer countries in terms of capacity-addition are also those with the largest domestic industries. National companies control their home base, affording opportunities for large-scale export. In turn, the configuration of the wind industry has impacted on national political sensitivities and the social acceptability of wind power in the latecomer countries, with probing questions being asked about the future of a sector which is both dependent on subsidies yet reliant on imports.

Will distinctive national models of wind power development be rendered obsolete by international convergence? The 'Danish model' was characterised by small-scale capitalism and local ownership (with utilities in the background), whilst the 'Spanish model' is still characterised by large-scale capitalism and national ownership (with utilities in the foreground). But in other countries, the 'international utility model' involving large-scale capitalism and international ownership is taking hold, due mainly to upscaling in the technology, but also encouraged by offshore operations. Although initial development is sometimes undertaken by medium-sized firms, the latter sell projects on to large corporations seeking to expand their energy portfolios internationally. The emergence of global players on the demand side (such as Iberdrola, Babcock Brown), mirrors developments in the supply side in turbine manufacture. The creation of global markets has accelerated trends to international consolidation as energy majors buy up stakes, with Shell, Areva, Siemens and General Electric already in the frame. These developments belie the 'alternative technology' ideology still propagated in relation to wind power by partisans. Thus the 'hegemonic battle' (Elliott, 2003: 185) is changing its contours – rather than just being a competition between 'old' and 'new' energy sources, or between 'conventional' and 'clean' energy purveyors, the question now is whether established energy firms will take over and/or drive out new entrants in renewables. In consequence, the wind sector is moving increasingly towards a 'bulk power', large-scale capital model, rather than to the small-scale, embedded generation 'Danish model', with its high level of community involvement and social acceptance.

Given the trend to 'large-scale penetration', the discussion of the integration of wind power into national ESIs in Chapter 6 identified important challenges in terms of generation mix and grid integration. Understanding of the 'fit' between wind power and the rest of electricity system is improved by comparing the generation mix across sample countries. Because of the need for back-up, wind power meshes well with systems having rapid response facilities, of which hydro is generally the best. However, hydro is not generally and uniformly available. On the other hand, wind has greatest environmental benefits when it substitutes for coal. Integration of wind power into systems with a high level of nuclear is problematic. Nuclear power serves to produce constant base load. But it is relatively inflexible, so cannot be ramped up and down rapidly, as can hydro or gas. However, wind power cannot replace nuclear power as base load, due to it intermittency (understood as unpredicted and uncontrolled generation). Thus the scenario of 'hello wind, goodbye nuclear' promoted by some greens is implausible. If base load does not come from nuclear power, with current technologies it must come from fossil fuel. Some countries use a combination of nuclear and coal for base load, but suggestions of the replacement of *both* by wind are fanciful. Because of these factors, the best structural fit for wind is found in Denmark, which has a coal-based ESI with no nuclear power and can balance its system across international connectors (allowing access to Scandinavian hydro). The worst structural fit is in France, where nuclear power provides over three-quarters of generation. This does not mean that wind power can have no place within it. Rather that place is far more constrained. Countries with a diversified generation mix, such as Germany, Spain and the UK, have greater scope for substitution of fossil fuel by wind, but its extent is a matter for 'learning by doing' over the medium term.

Saturation effects within the electricity system arise sooner or later, resulting in the grid management problems already experienced in Germany and Spain. The UK as an island system (or more precisely a collection of island systems since interconnectors between its constituents are currently limited) faces particular challenges. For the future, it will be important to identify the point at which total system costs (investment costs in generation, grid reinforcement and grid security) become unacceptably high. This is crucial since, if wind power maintains its current rate of expansion, grid reconfiguration will be inevitable in the medium to long term. Yet an increase in 'headroom' within national grids will ease but not resolve the operational difficulties associated with an 'atypical' generation source. In order to reduce intermittency problems, large-scale wind power will require management systems which draw closer to conventional dispatching, involving centralised control. The scenario of a concentration of generation conducted under national control procedures will decrease the local, 'embedded generation' aspect of wind power. Such a scenario also sharpens the question of the market integration of wind power, which is currently at a low level in several countries. If aspirational targets of 20–30 per cent of electricity from renewables are approached and met in at least some of the sample countries, and with wind being a major component, it is unlikely that such a large proportion of generation can lie permanently outside of market structures. Improved market integration will require extensive reform of policy instruments currently used to direct subsidies to the wind sector.

Policy learning

Sabatier (1993: 19) defined policy learning as 'relatively enduring alterations of thought or behavioural intentions that result from experience and are concerned with the attainment (or revision) of policy objectives'. Thus policy learning covers both incremental improvements in policy instruments and settings, as well as 'deep' changes in aims, content and procedures. In the case of the pioneer countries, policy learning has been on-going over more than two decades, offering opportunities for 'learning by observation' in latecomer countries.

Policy analysis in Chapters 4 and 5 revealed that conflict between categories of electricity provider was a core reason explaining why the deployment of wind power necessitated the establishment of 'political markets'. Established electricity purveyors had little interest or incentive in the development of wind power and, to varying extents, resisted its rise since it posed unwelcome costs. Subsidy schemes were introduced both to 'top up' returns on an emergent but as yet uncompetitive technology and to smooth out intra-sectoral conflict. National solutions proved distinctive. In Denmark and Germany, policy support encouraged the rise of new entrants and small-scale generation. The resistance of the utilities increased once this 'alternative' sector threatened the competitiveness of conventional generation. In Spain and the UK, however, policy support went mainly to utilities and large firms, with few new entrants and small-scale generation proving marginal. This resolved intra-sectoral tensions but at high costs. An 'alternative' set of energy purveyors did not emerge. Further, apparently 'market oriented' schemes proved more expensive, mainly because of embedded inflation pegs. With both the UK RO and the Spanish 'market option', wind power prices tracked the escalation of wholesale electricity prices over 2005–6. That escalation was mainly due to the increased price of oil and gas. Meanwhile subsidies to wind generation remained much the same, resulting in 'wind fall' profits. The problem of inflation pegs has therefore to be addressed by revision of the support mechanism.

This illustration is one of many examples of policy learning over two decades which have revealed substantial differences in levels of effectiveness, efficiency and equity of the various support schemes. Valuable lessons can therefore be learnt in order to improve current practice. In the Danish and German cases, feed-in tariffs proved highly effective in both increasing supply and creating a new category of electricity purveyor. Market entry by 'private' investors – whether as individual turbine owners or cooperative ownership of wind farms – provided a counter-balance to the power of the utilities. Because feed-in tariffs offered a low-risk investment environment and 'private' investors often had modest expectations regarding return on capital, tariffs were set relatively low (and reduced over time) yet still stimulated investment, leading to an efficient support system. This contrasts with the RO in the UK. Being risk-laden, it offers higher subsidies. The risks deter 'private' investors, but the corporations setting up or buying into large wind power stations have the capacity to manage the risk, making handsome returns from consumer subsidies.

With investment coming from distant third parties and profits returning to them, local communities receive minimal rewards yet carry the burdens. This contrasts with the Danish context which, as noted by Hvelplund (2001c: 21), provided a measure of equity through a local redistribution of gains, leading to a form of 'social contract':

> People like wind turbines when they own them and are not annoyed by the noise and visual inconveniences, especially when getting fair compensation. However, with a system of distant utility or shareholder owners, the local inhabitants get only the disadvantages and no compensation. This is seen as unjust and increases local political resistance to wind power.

Recognition of the impacts on communities is the initial step to improving the distribution of costs and benefits. A lowering of social and environmental costs is one route to increased acceptability, whilst a second is to increase benefits. Community trust funds are one means to achieve the latter but are far from constituting a panacea, given that the redistribution of consumer subsidies by large companies promoting their own interests raises ethical issues. Improved targeting of subsidies can realise efficiency gains and lead to improvements in equity for the wider public. For example, a lowering of subsidies to large corporations would reduce regressive burdens on consumers. This is well-understood in the Danish and German contexts, where on the one hand, the utilities have not been given access to wind subsidies, whilst on the other, tariffs to the 'private' sector have been reduced over time. Similarly, government proposals were made in Spain over 2006–7 to improve efficiency by the capping of subsidies, so as to help resolve the problems of indexing wind power prices on inflated gas prices. Yet in early 2007, UK policy-makers had still not reacted to comparable inflationary developments which were rewarding utilities whilst penalising customers.

These developments point up the permanent need to ask three questions in relation to support mechanisms, namely (1) whether installations need a subsidy at all, if so (2) what is the appropriate level, and (3) for how long is it needed. To establish whether a subsidy is needed requires transparency regarding costs and returns on the part of claimants. Companies have resisted transparency, however, pretexting commercial confidentiality. Consequently, opacity continues to be a problem, leading to contradictory statements being made about the need for subsidies. In its 1997 white paper, the European Commission (1997: 29) claimed in relation to its target of 40,000 MW of wind power in EU-15 by 2010, that 10,000 MW of offshore wind power would require subsidies but that 'no public financing will be needed for the 30,000 MW remaining installed capacity provided that a fair access to the European grids for the wind turbines is guaranteed'. Whilst access to grids has been provided by different procedures – guaranteed in the case of German and French feed-in tariffs, negotiated in the Spanish and UK contexts – consumer subsidies have been the reality everywhere. Subsidies were justified on the basis that the technology was 'emergent' and the sector was fragile. Yet the sector has made considerable progress to market maturity, raising the question of the continued need for subsidy. Based on the most recent information, using a 5 per cent discount rate and an average build cost of €1,000,000/MW, Milborrow (2007: 49) calculated that 'wind generation costs on a site with good winds of eight metres per second (m/s) can be as little as €42/MWh, rising to €52.5/MWh at 7 m/s and €71/MWh at 6 m/s'. With a higher discount rate of 8 per cent and a build cost of €1,400,000/MW, generation costs came to €68/MWh at 8 m/s and €85/MWh at 7 m/s. These levels offer a *rough* measure of overlap with subsidised prices in Germany and France, but are well below prices in the UK, at over €100/MWh. The scope to make large profits on windy sites explains why large investors have been rushing to increase their wind portfolios. There data reveal the importance of 'getting the prices right'. Regrettably, in a number of cases subsidies have been too high. Examples were identified in earlier chapters of continental policy-makers moving to correct the over-shoot, but none exist so far in the UK.

In order to move closer to the goal of 'fair and efficient' tariffs (Chabot, 2001), the longitudinal, cross-national comparison undertaken here indicates that an optimised support system for RES-E should be based on the principles of technology differentiation and cost reflective subsidies in relation to each conversion technology and its contexts of usage. This requires explicit statements over why subsidy is necessary and at what level, in relation to a specified quantity and quality of generation. The circumstances under which the support will be continued, reduced or phased out should also be specified. A key recommendation is that the level for capping be defined in terms of the relationship between the average wholesale price of electricity (calculated yearly) and the real cost of RES-E. Where RES-E generation from a particular source costs more than the wholesale price, an 'environmental premium' is worth paying in specified circumstances. But where the wholesale price systematically exceeds RES-E generation costs (calculated to include a fair return on investment), subsidy should be stopped with rapid effect.

The design of future support systems must also take into account perverse outcomes encountered by current schemes. Given the international mobility of capital and the shortage of manufacturing capacity for wind technologies, in 2006–7 markets were stood on their head. In a sellers' market, wind turbine manufacturers faced few competitive pressures and increased prices substantially (though this was partly in response to price rises for raw materials). Meanwhile national support schemes bidding for wind turbines to meet RES-E targets for 2010 found themselves in competition *with each other*. This outcome suggests a case for more European coordination, not necessarily in terms of a common *choice* of instrument but in agreements regarding their *settings*, such that cross-border 'cherry picking' of subsidies by international operators is minimised. It also confirms the advisability of *not* setting compulsory capacity targets in contexts of manufacturing shortages or of oligopoly supply.

The problem of opacity also arises in relation to emissions reductions achieved by wind power deployment. There is no commonly validated system for the calculation of indirect emissions savings achieved through the displacement of fossil fuel by wind power. This unsatisfactory situation must be resolved through the establishment of a scientifically robust, testable and replicable methodology agreed by independent experts. In addition, now that the institutionalisation of carbon trading has created a new category of assets with a potential market value of billions of euros, transparent and rigorous *auditing* of both emissions and emission savings is as essential for carbon as it is for any other category of assets or liabilities. The methodology should result in accurate, meaningful and easily accessible data on emission baselines at suitably disaggregated levels, in order to identify the direction of emission trends and determine whether progress is in terms of 'avoidance' or actual reductions. The impacts of new climate policy instruments, such as EU-ETS, also need to be monitored in terms of their impacts on RES-E. These new instruments are likely to accelerate price convergence between fossil fuel and renewable energy sources, constituting a further reason to reform or indeed abolish subsidies. There is, after all, no reason why the 'greenness' of renewables should be paid for twice. Consequently, it is recommended that regulators be alert to 'green power' schemes which, in charging a premium to customers whilst drawing a mandatory subsidy, may be selling their 'greenness' twice.

However, lucidity is required over interactions between short- and long-term policy goals. The major instruments used to support wind power have been investment subsidies and production subsidies. The immediate aim was to increase generation from a renewable energy source. In continental systems, broader economic goals have also been fulfilled. Crucially, REFITs have proved successful in serving both the *energy policy* purposes of diversified generation *via* RES-E and the *industrial policy* purposes of promoting the manufacturing sector. In the pioneer countries, higher energy costs have been offset by technological leadership, employment creation and export opportunities in a new industry. But emulation has not proved straightforward. In France, the need for industrial policy to develop renewables was recognised in the parliamentary report by Birraux and Le Déaut (2001: 261). In the UK, the DTI (2004: 1) described renewable energy as 'a huge opportunity to enhance our manufacturing capacity and provide new employment'. However, the history of wind power has demonstrated that the development of RES-E conversion technologies is heavily dependent on national industrial structures and technology-forcing policy. Germany, in particular, benefited from an existing manufacturing base in electrical engineering which was stimulated by subsidies to wind power and, through the VDMA, turned into a powerful voice lobbying for their continuation. No comparable dynamic can be identified in the UK where the RO serves the narrow purpose of pulling through 'lowest cost technologies sequentially' (Carbon Trust, 2006: 2). In rewarding technologies that are already close to the market, the RO neglects the long-term aims of proactive industrial policy. Hence Ofgem (2007: 1) commented 'there is little evidence so far that the RO is encouraging technological development'. Marine renewables have been identified as an important sector for securing the energy needs of the UK, but the industrial policy frame required to move from optimistic aspirations to manufacturing capability has yet to be identified. Whilst new research and policy learning will be required to remedy this, a core lesson gained from the wind power context is recognition of the inability of the RO to meet industrial policy needs. Thus whilst adjustments in its settings (namely, to lower subsidies for windy sites) would lead to greater efficiency in the short term, its inefficacy for the long term necessitates a search for alternatives.

Institutional challenges arising from the planning system were identified in Chapter 7. In the pioneer countries the combination of indicative planning (focusing on capacity targets) and spatial planning (focusing on strategic locational guidance) assisted policy implementation, by providing a 'joined up' response that linked policy intentions with deployment on the ground. France is moving over to this approach with the establishment of 'wind power development zones'. Exclusion/inclusion zones reduce uncertainties for interested parties, and may lead to a reduction in transaction costs. Where they are drawn up on the basis of consultation and participation, they form a variety of 'social contract' into which local communities may enter consensually, presumably because local/regional benefits are forthcoming. In the UK, however, the planning process relies on a criteria-based style of decision-making, involving significant uncertainty over outcomes. The planning reform undertaken by the ODPM in the early 2000s, notably with the establishment of PPS22, appears to go in the direction of greater centralised control, as does the trend to executive intervention under section 36 of the 1989 Electricity Act. A 'top down' decision-making style and central locus of control help explain the conflicts arising in UK over wind farm proposals. Conversely, the social dynamics contributing to acceptance of wind power in the pioneer countries – societal participation in strategic locational guidance, community ownership and wider socio-economic advantages – are largely missing in the UK and proving difficult to foster. For the future, the problems of saturation (namely, using up accessible windy sites) and increased cumulative effect require the attention of policy-makers in the pioneer countries and indeed in some regions within the latecomer countries. Further, as indicated in Chapter 6, the lack of grid connections is as much – if not more – a cause of delay, than getting planning consents.

The difficulties in resolving wind power policy and planning issues have been compounded by the bounded mindset of policy-makers and by partisan discourses. A blindness to the wider implications of the integration of renewables was already evidenced in the white paper by the European Commission (1997: 34–6) whose 'action plan' was limited to technocratic, top-down measures which targeted industrial and technology development, but made no mention of public consultation or involvement other than in terms of 'consumer information campaigns'. This was perhaps excusable in the 1990s, given limited experience with renewables. The problem is that ten years later the same mindset persists. A frequent assumption is that renewable sources of generation can simply slot into the place of conventional resources, whilst leaving other economic and social aspects of electricity generation, distribution and use unchanged. Perhaps the biggest lesson to be drawn from this five country survey of wind power is that this assumption is incorrect. Questions of production cannot be left up to the large utilities. Questions of consumption must include demand management and reduction. Both the supply and demand sides of the equation require the active involvement of a wide range of economic and social actors, rather than passive 'information campaigns'. Recourse to renewables is already involving major changes not just in generation technology, but in terms of social and environmental impacts. The latter arise *inter alia* from locational issues and from socio-economic impacts in terms of the distribution of costs, given the regressive effects of subsidies financed through consumer bills.

Keeping the public out of the debate is no longer an option, now that a range of organisations, including public bodies, long-standing NGOs as well as new antiwind groups, have demanded that their views be heard. The involvement of a wider circle of actors in energy and climate policy debates is to welcomed, whilst a re-balancing of the relative influence of big business lobbies and civil society organisations is arguably to be encouraged. Yet participative procedures have barely evolved, since policy-making has concentrated on the technical dimensions of economic support mechanisms with limited attention to broader issues.

Cross-national comparison of the contingent approaches to the promotion of wind power has revealed the hollowness of partisan arguments which encourage categoric stances for *or* against wind. Analysis in Chapter 3 of the discourses of the wind lobby revealed how a heterogeneous coalition of industrialists, environmentalist NGOs and politicians formed to promote wind power on the basis of 'story lines' stressing the relations between energy supply and climate change. Their 'story lines' provided the cognitive and communicative conditions to rally around the slogans of 'clean' energy and voice a resounding 'yes to wind'. Yet the use of wind energy to generate electricity is not a categorical imperative, with black and white moral entailments. It is merely one solution of convenience among a range of conversion technologies, energy conservation measures and demand management techniques. Because the manners of promotion of wind power entail consequences in terms of effectiveness, efficiency and equity, the *conditions* of its deployment are paramount. These conditions vary, and can be acted upon for better or for worse.

Chapter 8 identified a range of established organisations that contested wind power on the basis of conditional objections which also enunciated acceptability criteria. Landscape and countryside organisations have sought to define what is 'environmentally acceptable' by calling for wind farm construction to be excluded from such sensitive areas as national parks, 'emblematic' landscapes and heritage coasts. The ornithological associations have called for strict protection of endangered species and habitats designated under European and international conventions. To help resolve the tension between protection of natural heritage and its acceptance of renewables, SNH (2005a) prepared 'strategic locational guidance' for wind farms in Scotland. Likewise CPRE (2006) sought 'a sequential approach (...) to steer wind development to the least environmentally sensitive areas'.

This approach based on conditionality contrasts with the stance of antiwind groups, whose opposition can appear generic. Local wind protest groups seek denial of planning consent to a specific wind farm application, but their attitude to deployment in their immediate vicinity may or may not involve a blanket rejection of the technology. The stance of some antiwind protesters clearly reflects the conditionality approach of established conservation NGOs; indeed, there are individuals who belong to both categories of organisation. However, the attitude of the leaders of certain antiwind umbrella organisations is characterised by systematic opposition. This has created ambiguity and uncertainty over the wider goals of antiwind groups, allowing prowind partisans to tar all such protest with the same brush. The clash of partisan perspectives has ignited media interest. Controversy has served to alert the public to the issues, piqued their curiosity and catalysed engagement, but displays of passion should not shrink the will for informed inquiry.

Unfolding path choices

Baumgartner and Midttun (1987b: 291) proposed that 'the role of the social sciences (...) is to make people aware of options and alternative paths of development'. Accordingly, Chapter 1 unfolded three categories of choice related to renewables. The first of these was *socio-economic* choice, with an alternative between what Lovins called a 'hard path' and a 'soft path'. The second was in terms of *policy frame* choice, with the stress falling either on the economic frame or the environmental frame. The third related to the *governance* choice embedded in the differences between 'sustainable development' and 'ecological modernisation'.

In Chapter 2 it was shown that the ideological values of the alternative energy movement – environmentally benign sourcing, local participation, embedded generation, decentralised consumption – have continued to be rallying points for activists who promote wind power, The 'Danish model' of the 1980s proved closest to these ideals. But the problem is that, with the exception of neighbouring German *Länder*, the 'Danish model' has not spread to other countries. On the contrary, in response to the ambition to achieve 'large-scale penetration' of ESIs by wind power, the 'international utility model' has taken off in Europe and is spreading globally. The trend to 'super-size' the wind sector – namely, the upscaling of turbines, the spread of large wind power stations and the pressures for regional concentration – has encouraged dominance by international corporations because of the extensive capital requirements inherent in the large-scale deployment model. But whereas small installations under community ownership involved a 'soft' path, 'super-sizing' puts the wind sector back on a 'hard' path. Currently, it is uncertain whether the 'soft' path has been entirely superseded, since local ownership continues to be a reality in northern Europe, a new generation of small turbines is reconfiguring the feasibility of 'micro-generation', and the balance between generation sources cannot be predicted for the long term. But what is clear is that in the 2000s the international utilities and affiliated corporations staged a takeover of the wind industry, seeking to develop a 'bulk power' variant where generation takes place mostly on the geographical periphery but is centralised in terms of industry structures, ownership and management. Thus was done under the cover afforded by the ideals of the alternative energy movement and environmentalist NGOs, and on the back of substantial consumer subsidies.

The *socio-economic* choice taken by government and big business in favour of a 'hard path' impacts on *policy frame* choices. The promotion of wind power has typically stressed the environmental benefits of 'clean' energy resulting from zero emissions at the point of generation. However, due to anxieties over the future sourcing of oil and gas from politically unstable or unreliable regimes, classic economic arguments related to 'security of supply' and 'energy independence' resurfaced in the mid-2000s. This tipped prowind discourse towards a stress on indigenous sourcing, and on the absence of problematic fuel supply infrastructure (no quarrying, no pipelines, and so forth) in the case of wind power. At the same time, the emphasis on energy security arguments reconnects with the traditional goal of economic policy, namely the promotion of growth. Thereby wind power becomes part of the bulk energy base on which economic expansion is premised.

The combination of bulk power, hard path and economic expansion add up to a programme of industrial restructuring that can be summarised as an 'ecomodernist growth paradigm'. The wind power sector is an important and revealing example of the twin faces of ecological modernisation. As sociological theory ecological modernisation stresses the compatibility between high environmental performance and high economic performance, but as practice it shows how environmental challenges create opportunities for the reinforcement of international capitalism. On the other hand, large-scale deployment of wind power has little in common with the sustainable development paradigm for two reasons. One is that sustainable development involves attention to three pillars – the economic, the environmental and the social – but the social pillar is largely lacking in the wind sector's current development trajectory. The other relates to ambiguities within the notion of 'sustainability'. Wind energy is inexhaustible, but the sustainability of an energy source does not guarantee the sustainability of the economic model its supports. The irony of the 'clean energy' story line is that its promise of modernisation has been used to legitimise 'business-as-usual' practices. More specifically, unsustainable levels of consumption are now being justified on the basis of the sustainability of renewable energy sources. However, the capacity of renewables to meet the purposes of economic expansion is a separate issue to their inherent sustainability: the latter is *not* a guarantor of the former. This raises the question of whether the 'ecomodernist growth paradigm' will prove feasible in the long term.

In summary, inter-relationships have been identified between the three categories of development path choices. A 'hard path' unfolds before the wind sector due to the stress on 'large-scale penetration' and bulk power, provided by large installations and major corporations. This path leads to predominance of the economic policy frame, with enhanced environmental performance serving as a precondition for continued expansion. It also predisposes in favour of 'ecological modernisation' and the influence of economic actors, whilst downplaying the societal aspirations contained in the sustainable development paradigm. Thus in relation to the question asked in Chapter 1 regarding how we move up the ladder of decisional complexity that is formed by the three categories of choice – namely, how we move from socio-economic choice, to policy frame choice and on to governance choice – the answer is that choices taken in relation to lower rungs determine higher level choices, and *not* vice versa. Perhaps unsurprisingly, the macro-realm of governance is contingent on and emerges from micro-level decisions. But this insight can be used to fashion future development paths.

Towards a social contract for renewables

Chapter 1 introduced the idea of a 'social contract' between energy industries and society by which Haugland, Bergesen and Roland (1998: 18–19) understood a 'quid pro quo' whereby 'actors in the sector are given societal objectives to which they are committed – security of supply, employment, environmental goals, etc. In return, they acquire a carefully defined freedom of action'. Defining this 'freedom of action' involves a process of societal negotiation, but one in which economic actors have traditionally held the upper hand. A social contract for renewables involves redressing the balance and opening a space for re-negotiation. Focusing on the micro-level of socio-economic choices, such a social contract would start with the capacity of local communities to identify their energy needs and to recombine elements of 'hard' and 'soft' paths to respond to those needs, in line with evolving energy sourcing and conversion technology developments. The contract cannot be purely local – it must also have national, European and international elements. But to understand, implement and have influence over national and supra-national energy and climate policy, citizens need to orientate the local sphere – individually, as households and collectively.

At the heart of any contract is the understanding that acceptance is as much *process* as it is *product*. Acceptance arises on the basis of meeting criteria and conditions. Thus acceptance and rejection are not 'one off' decisions, taken for ever. As acceptability criteria and conditions evolve, so too do social responses. This study has identified a range of conditions and contingencies that reflect the diversity of contexts in which wind power has been inserted. With regard to renewables in the round, the assumption is therefore that in the future many permutations will be possible, based on need, constraint, opportunity and preference. Thus it is not possible – nor desirable – to predefine the contents of any particular social contract for renewables. But some generic criteria and principles can be indicated here, whilst others must form the remit of further research. They include:

- equity – a fair distribution of burdens and benefits;
- transparency – of costs, prices, subsidies and profits;

- openness – of membership of decision-making bodies and of their deliberations;
- scrutiny – by legitimate and concerned parties;
- accountability – in relation to personnel and procedures;
- review – to allow social learning and to act upon it.

Social contracts for renewables, elaborated on the basis of societal negotiation using explicit acceptability criteria, can make a practical contribution to sustainable development agendas. As such, they could act as a counterweight to the tendency within the practice of 'ecological modernisation' to prioritise the preferences of a limited set of economic actors.

The politics of renewables: pathway options and unresolved challenges

Wind power is proving to be a 'path finding' technology not only in terms of the 'solutions' it provides, but also in terms of the unresolved issues it raises for renewables and for energy sourcing more widely. Four major challenges can be identified:

1. In relation to renewables, how far is policy to be state-directed or market-centred?
2. In relation to energy sources generally, will fuel competition or political arbitration decide development trajectories?
3. Will preference go to bulk power or decentralised production?
4. Will patterns of governance arise from and encourage 'bottom up' societal engagement or 'top-down' technocratic management?

In relation to the deployment of large wind turbines, answers have now been given to these questions for the historical period covered. But in relation to other technologies, the answers will be more open-ended, particularly as we look further into the future. Raising these questions at this juncture serves the purpose of identifying the main political dimensions with which 'social contracts for energy' will need to engage.

Table 9.1 summarises the political challenges and their implications. It lists four sets of alternatives and links these to different layers of decision-making in the polity. The first alternative is whether renewables policy will be state-directed or market-led. The main choices relate to the selection of policy instruments and financial support schemes. The political domain is that of 'low politics' but, in an era of

Table 9.1 Summarising the political challenges of energy sourcing

Challenges	Choices	Domain	Salience
1. state or market?	policy instruments; support schemes	low politics	high
2. fuel competition or political arbitration?	upstream 'formatting' of the energy sector	high politics	relatively high
3. bulk power or decentralised supply?	security, independence, equity	public goods	relatively low
4. societal engagement or technocratic management?	institutions; values	governance	low

rampant neo-liberalism, the salience of the alternative has been high in practitioner debates and in the academic literature. The second alternative is whether the 'battle' between energy sources will be resolved by fuel competition in the market or political arbitration by governments. The choices relate to the 'upstream' aspects of government-big business relationships, including the role of international diplomacy in 'formatting' oil and gas markets, political negotiations over the future of nuclear power and so forth. As such, they form part of 'high politics', with relatively high salience in periods of 'business as usual', rising to high at times of crisis and shortage. The third alternative is whether decentralised supply will make a come-back against bulk power. Although the trend to large wind power stations has provided one answer, it has also made the question more prominent in relation to other renewables. These questions are particularly acute in RES-E because of the economic, social and environmental costs of reconfiguring transmission and distribution grids. The choices here of different orders, including security of supply and equity. The political domain is the production of 'public goods'. Its salience is relatively low, in that security of supply issues had a low profile in the 1990s but their profile was rising in the mid-2000s. Further, the salience of equity issues has also been low but rising, now that liberalisation of electricity and gas markets is a reality for households and equity provisions contained in 'social contracts' from the era of monopoly suppliers are dissolving. The fourth alternative is whether societal engagement or technocratic management will prevail. The choices relate to the structure and functioning of institutions, and the values they embody. Thus the political domain is the evolution of governance systems. The salience of these issues has been low, but is expected to increase.

In reverse sequence, these four alternatives connect to abstract democratic ideals such as participation (4) and justice (3), and to their translation into practice *via* economic processes of distribution (2) and redistribution (1) of material benefit. Indeed, it is by going through this reverse sequence that the limits of the current policy frame towards renewables are made stark. The debate has often fastened on the narrow dimension of policy instruments, and either avoided the broader political dimensions altogether or else given simplistic responses.

Thus the four challenges serve to map the contours of the politics of renewables and reveal its extent and significance. This conceptualisation provides opportunities for political science research. The renewable energy domain is more than just an energy or technology choice, but also a laboratory whose experiments are emblematic for the sustainability transition – with its promises and pitfalls. Finally, although strong elements of path dependence have been noted in wind power development – especially liberalisation as a source of greater market and political power for international energy corporations – the scope for social learning and democratic reform has also been stressed. It is hoped that this conceptualisation of the issues will contribute to identifying new research agendas, to improve understanding of the problems we face, and to feed into the policy responses to tackle them.

Concluding remarks

The advancement of knowledge involves answering research questions, but the process always throws up new questions and serves to stimulate further debate. How are we to choose between development path options? Can we simply give an opinion, or do we need further social learning to make informed and viable choices? What societal processes and institutional arrangements are required for the drawing up of 'social contracts' in the energy sector? Indeed, will there be debate, deliberation and collective negotiation, or simply a top-down imposition of decisions? In seeking to convert from finite conventional energy sources to sustainable renewables, we are at the start of a learning curve with little idea of where it will take us. On the one hand, this can be viewed as productive in that exciting, creative solutions can be developed. On the other hand, powerful economic and political actors are already working to ensure that the development paths taken will suit their interests.

Yet the public is not bereft of opportunities for action. Relatively simple demands can have multiple repercussions. If the public required that consumer subsidies went to a wide range of energy-related technologies, industries and companies, and that a bigger share of the 'public pot' was returned equitably to households and communities for discretionary spending on renewable technologies *and* energy savings, what impact would this have on development pathways? And in what ways would the outcomes contrast in social, economic and political terms with a bulk power, big business future? An ambitious public could even demand the reconfiguration of the electricity system. This would involve the capping of large-scale generation and usage as a key near-term milestone. The recovery and use of heat from electricity generation processes, the increase in on-site production, the reduction of long-distance transmission losses are longer-term goals that would help solve energy and climate policy equations. Successful demand-side management would, over time, improve efficiency, downsize aggregate generation capacity and output, decrease investment needs and further reduce environmental harm. And most importantly, as part of a new 'social contract' the public can demand that the benefits of improved efficiency and renewable energy flow back to them. Indeed, cutting escalating energy bills is one strong reason for making these demands; another is making a contribution to climate protection that matters. In the energy sector as elsewhere, an active and involved citizenry is perhaps the only way to make sustainable development a reality.

Notes

1 Contextualising the Wind Power Debate

1 Quoted from the *Sunday Telegraph*, 18 April 2004, p. 3. Sir Jonathan Porritt was formerly Director of UK Friends of the Earth, and now chair of the Sustainable Development Commission. Sir Bernard Ingham was chief press secretary to Mrs Thatcher's government, and is now a vice-president of Country Guardian, an antiwind association.
2 See Krewitt (2002) and Sundquist (2004).
3 The 'sustainable development' concept has an extensive history, but the Brundtland Report is often treated as a seminal source. Useful commentaries include Redclift (1987) and Dobson (1998); collections of articles can be found in Baker *et al.* (1997), Barry, Baxter and Dunphy (2004), Kirkby, O'Keefe and Timberlake (1995), O'Riordan and Voisey (1998). A contrarian opinion can be found in Beckerman (1995). For recent re-appraisals, see Meadowcroft (2000) and Carruthers (2005).
4 See Weizsäcker, Lovins and Lovins (1997).
5 See Toke and Strachan (2006).
6 See Luckin (1990).
7 See Szarka (2004, 2006, 2007).
8 See Szarka and Blühdorn (2006).

2 Diagnosing the Wind Sector

1 This table is reproduced verbatim. Percentages do not sum to a hundred because of data reconciliation problems acknowledged in the source text.
2 For the history of the Danish wind turbine sector see Farstad and Ward (1984), Van Est (1999: 69–96) and Bergek and Jacobsson (2003).
3 See www.windsupply.co.uk.
4 For example, Quebec imposed a 30 per cent local content requirement (Bailey, 2005: 64) whilst China demanded 70 per cent (Jianxiang, 2006: 70).
5 For more extensive discussion, see Flam (1994), Kitschelt (1986), Wurzel (2002), and Blühdorn and Szarka (2004).
6 See Edwards (1994).
7 See for example Rand and Clarke (1990), Hinshelwood *et al.* (2000), Toke (2005c).
8 The practice of 'repowering' – namely, the replacement of old, small turbines with large, new ones – takes advantage of these factors.
9 For example, Cefn Croes held the record in 2005 at 58.5 MW, then Black Law at 97 MW in 2006, followed by Hadyard Hill at 120 MW, whilst Whitelee at 322 MW was consented in 2006.
10 Sources of information on offshore costs include Ecofys (2002: 51); Sustainable Development Commission (2005: 29) and Massy (2005).

3 Mobilising for Wind Power

1 For full names of these organisations and their translations, please see list of abbreviations.
2 For discussion of the VDMA within the German wind lobby, see Michaelowa (2005).
3 For details, see Mez and Piening (2002)
4 For example, there is an EU R&D programme on 'intelligent energy'.
5 The question of methodologies for calculating emission reductions will be held over till Chapter 6.
6 Quoted in Anderson and Bailey (2006: 52).
7 See EWEA (2006).
8 See Luckin (1990: 9–22) on the persistent 'triumphalism' and 'evangelicism' of the electricity industry.

4 Promoting Wind Power through National Policies

1 In the 2000s, pound to euro values have fluctuated mostly in the range of 1.4 to 1.5.
2 For discussion of the making of the directive, see Rowlands (2005). For EU renewables policy, see Lauber (2005a).
3 Useful overviews can be found in Gipe (1995), Redlinger, Dannemand and Morthorst (2002), Sawin, (2004) and IEA (2005a).
4 A typical exchange rate was 1 euro = 7.45 DKK.
5 According to Agnolucci (2006), production incentives reached €90 million in 1998.
6 For discussion of this concept, see Héritier, Knill and Mingers (1996).
7 For details of tariffs, see the Danish Energy Authority website: http://www.ens.dk.
8 In the early 2000s, repowering led to the removal of approximately 900 older turbines (IEA, 2005a: 84).
9 See Bergek and Jacobsson (2003) for analysis of the development of the German wind turbine industry.
10 A recent estimate of these was 2.4 €c/kWh (Eurelectric, 2004: 11).
11 There are four TSOs in Germany: RWE (west), EnBW (south-west), E.ON Netz (north to south-east corridor) and Vattenfall (East). As of April 2003, E.ON Netz had 5,500 MW of wind power in its catchment (Luther, Radtke and Winter, 2005: 233).
12 The *régimen especial* has a regulatory framework which is distinct from the *régimen ordinario* for conventional sources.
13 Data from http://www.suivi-eolien.com.
14 Offshore tariffs were also set, at 13 c€/kWh for ten years, falling to between 13 and 3 c€/kWh (depending on number of full load hours of operation) for a further ten years.
15 See Elliott (1992: 258–9).
16 For discussion of the NFFO and UK renewables policies, see Mitchell (2000), and Mitchell and Connor (2004).

17 In addition, for industrial users the value of RES-E is increased by 0.43 p/kWh, which is the level of the Climate Change Levy (CCL), introduced in April 2001. The CCL is raised on electricity purchased by industry from all conventional (including nuclear) sources, but RES-E is exempt.

18 The 'final decisions' taken concerned minor technicalities and can be found in DTI (2006a).

5   Drawing Policy Lessons from Cross-National Comparisons

1 See Baumgartner and Midttun (1987a).

2 Exceptions such as Baywind in Cumbria and the Westmill Wind Farm Cooperative in Oxfordshire tend to confirm the rule.

3 See Eurelectric (2004: 16–17).

4 Estimates of 10,000 MW can be found in Cochet (2000: 114) and Boston Consulting Group (2004: 15). Birraux and Le Déaut (2001: 368) proposed a higher estimate of 14,000 MW of capacity.

5 A more detailed version of the following discussion can be found in Szarka and Blühdorn (2006).

6 However, ROC prices in auction move up and down. In April 2006, the auction price was back to £40.65 (NFPA, 2006).

7 Exchange rates fluctuate, mostly in the range of 1.4 to 1.5 euros to the pound.

8 See Sustainable Development Commission (2005: 125).

9 The IEA (2006a: 19) noted that O & M costs were rarely reported, but gave 7 €/MWh as typical in the Spanish and Swedish cases.

10 Hvelplund (2001b: 61) uses the term 'energy automaton'.

11 See Tinkerman (2006).

12 Eligibility for the RO depends on whether an energy source and its usage is designated as 'renewable' by policy-makers. Questions of eligibility have proved tricky in some areas, for example the practice of 'co-firing' biomass in coal-fired power stations.

13 However, section 185 of the UK 2004 Energy Act gives discretion to the DTI over the setting of transmission charges in relation to outlying areas with significant RES-E potential, such as the Scottish islands.

14 Ofgem (2007: 17) provided a detailed list of reasons why extension beyond 2015 is unjustified.

15 For an insightful discussion, see Hvelplund (2005).

6   Integrating Wind Power into the Electricity Supply Industry

1 For details, see Mez and Piening (2002) and Wüstenhagen and Bilharz (2005).

2 For example, In Germany, the coal sector lobbied the SPD and mounted a 'virulent campaign' (Bechberger and Reiche, 2004: 55) against wind power in 2003, leading to Chancellor Schröder agreeing to subsidise hard coal-mining by 17 billion euros between 2006 and 2012.

3 See Midttun (1997: 283–5).

4 More precisely, recourse to 'balancing power' involves responding to unanticipated changes in demand – or non-delivery of scheduled generation – by calling on three categories of reserve: 'regulating reserve' comprised of plant to maintain correct frequency, 'spinning reserve' which is operational at less than full output but ready to be brought on line, and 'standing reserve' which is not generating but ready to do so (Milborrow and Harrison, 2004: 38). These facilities cope with short-falls in generation. In addition, if a higher quantity of wind power arrives than is expected, arrangements need to be in place to reduce output from a conventional generator. Thus 'balancing power' works in both directions, up and down.

5 The more accurate the weather forecasts, the easier it is to integrate wind generation. Jackson (2003) noted two main types of forecasting error: wind speed error and 'phase error'. Wind speed error results in deviations in output from forecast. With 'phase error', the wind speed prediction is correct but estimation of the timing of the weather front is incorrect. In the case of late arrival, this can lead to the TSO waiting for a wind front to arrive and being short of scheduled electricity in the meantime, so having to purchase more on balancing markets at a higher price. In the case of early arrival, the TSO will have non-scheduled, surplus wind generated electricity to dispose of, probably at a loss.

6 They can only do this to the extent that the regulatory framework will allow, and disputes arise over calculation methods. For example, in 2006 Vattenfall Europe Transmission was in contention with BNA, the Germany regulator, over its level of network and balancing charges to consumers leading to ongoing legal proceedings. See Knight (2006).

7 For details of the German offshore wind strategy, see BMU (2002) and Viertl and Bömer (2005).

8 They probably include: the problems of conceptualisation; the pace of change; limited incentives and finance for modelling work; inadequate auditing; data retention on the part of generating companies.

9 One tonne of carbon is equivalent to 3.66 tonnes of CO_2.

7 Siting, Planning and Acceptability

1 See Edwards (1994).

2 See Danish Ministry of Environment and Energy (1996: 41).

3 For detailed discussion of the German system, see Breukers (2006: 200–11).

4 For analysis of practices in the regions, see Blázquez; Calero de Hoces and Lehtinen (2003) on Andalusia, Calero and Carta (2004) on the Canary Islands, and Faulin *et al.* (2006) on Navarra.

5 See Scottish Executive (2000).

6 See Department of the Environment (1993).

7 The proposals for a wind farm at Whinash in the Lake District involved a small but real incursion into a national park. The proposals were rejected after a lengthy and onerous public inquiry. For the inspector's report, see Rose (2006).

8 For data, see Chapter 2.

8 Contesting Wind Power

1 For participant accounts of two *causes célèbres* of protest against wind power, see Mann (2003) on Barningham High Moor and Little (2003) on Cefn Croes.
2 See for example GURELUR (2006).
3 For discussion see Lindsay (2005).
4 See REF / Hall (2006).
5 See Country Guardian / Etherington (2006: 35–9).
6 For example, FoE (2004: 2) claimed that 'doubling nuclear power in the UK would reduce greenhouse gas emissions by no more than 8 per cent at most'.
7 For discussion of third-party rights in planning, see Ellis (2004).
8 See for example Mann (2003) and VoS (2003a: 1).
9 REF (2004a: 5) argued for 'a balanced approach (...) we need a "bit of everything" and this is especially true for renewables'.
10 In some cases, they have been forced to retract their claims: see VoS (2005: 24).
11 Compare the statements made by BWEA (2004) and their reprise by Yes2Wind (2004) and DTI (2006b).
12 The numbers of proposals not proceeding to an application are difficult to ascertain accurately. However, there are reasons to believe that they are substantial. For example, in 2004 and in relation to Scotland alone, the SNH (2004) inventoried 6721 MW of onshore wind proposals in the preapplication stage (including scoping), 2255 MW in the application stage and 918 MW approved.
13 Natural England is an agency which amalgamated English Nature, the Countryside Agency and the Rural Development Service.
14 For discussion, see RSPB and BirdLife International (2003).
15 See Countryside Agency (2003). For the inspector's report, see Rose (2006).
16 For a historical survey, see Luckin (1990: 156–71).
17 See for example Dalton (1993: 60–1), Norris (1997) and Witherspoon (1994).
18 See for example Mann (2003: 31) who emphasised that the fight against the proposed wind farm on Barningham High Moor was not 'antiwind' but 'pro-landscape protection'.

References

AEE (2006) *Wind Power 2006*, Madrid: AEE

Agnolucci, P. (2006) 'Factors influencing the likelihood of regulatory changes in renewable electricity policies', *Renewable and Sustainable Energy Reviews*, in press

Aguilar Fernández, S. (2003) 'El principio de integración medioambiental dentro de la Unión Europea: la imbricación entre integración y desarrollo sostenible', *Revista de sociología*, 71: 77–9

Alt, H. (2005) 'The economics of wind energy within the generation mix', *International Journal of Energy Technology and Policy*, 3: 1–2, 158–82

Anderson, M. (2006) 'High level action to solve radar issue', *Windpower Monthly*, 22: 7 (July), 26

Anderson, M. and Bailey, D. (2006) 'AWEA Conference Report: a tricky market of shifting dynamics', *Wind Power Monthly*, 22: 7 (July), 45–52

Anon (2006) 'Virginity lost', *Windpower Monthly*, 22: 7 (July), 17–18

Arts, B. (2002) 'Green alliances: of business and NGOs: new styles of self-regulation or dead-end roads?', *Corporate Social Responsibility and Environmental Management*, 9: 26–36

Aubrey, C. (2005) 'The Spanish wind market: dynamic and focused', *Wind Directions* (July/August), 16–20

Avia Avanda, F. and Cruz Cruz, I. (2000) 'Breezing ahead: the Spanish wind energy market', *Renewable Energy World*, 3 (May–June), 35–45

Badelin, A., Ensslin, C. and Hoppe-Kilpper, M. (2004) 'Does the wind blow stronger in Europe? Current experience with supporting wind power in European markets', Kassel: ISET, http://www.iset.uni-kassel.de/abt/FB-I/publication/04-03-28_awea.pdf Consulted 11.10.2004

Bailey, D. (2005) 'Handcuffed by local content demands', *Windpower Monthly*, 21: 10 (October), 64–5

Baker, S. *et al.* (eds) (1997) *The Politics of Sustainable Development. Theory, Policy and Practice within the European Union*, London: Routledge

Barry, J. (2005) 'Ecological modernisation', in Dryzek, J. *Debating The Earth: The Environmental Politics Reader*, pp. 303–21, Oxford: Oxford University Press

Barry, J., Baxter, B. and Dunphy, R. (eds) (2004) *Europe, Globalization and Sustainable Development*, London: Routledge

Bartle, I. (2004) 'Energy', in Compston, H. (ed.) *Handbook of Public Policy in Europe: Britain, France and Germany*, pp. 194–204, Basingstoke: Palgrave

Bataille, C. and Birraux, C. (2006) *Rapport sur les nouvelles technologies de l'énergie et la séquestration du dioxyde de carbone: aspects scientifiques et techniques*, Paris: Office parlementaire d'évaluation des choix scientifiques et technologiques / Assemblée nationale, rapport no. 2965

Baumgartner, T. and Midttun, A. (1987a) 'Energy forecasting and political structure: some comparative notes', in Baumgartner, T. and Midttun, A. (eds) *The Politics of Energy Forecasting: A Comparative Study of Energy Forecasting in Western Europe and North America*, pp. 267–89, Oxford: Oxford University Press

Baumgartner, T. and Midttun, A. (1987b) 'Modelling and forecasting in self-reactive policy contexts: some meta-methodological comments', in Baumgartner, T. and Midttun, A. (eds) *The Politics of Energy Forecasting: A Comparative Study of Energy Forecasting in Western Europe and North America*, pp. 290–308, Oxford: Oxford University Press

Bechberger, M. and Reiche, D. (2004) 'Renewable energy policy in Germany: pioneering and exemplary regulations', *Energy for Sustainable Development*, 8: 1 (March), 47–57

Beckerman, W. (1995) *Small is Stupid. Blowing the Whistle on the Greens*, London: Duckworth

Bell, D. R. (2004) 'Sustainability through democratization? The Aarhus Convention and the future of environmental decision making in Europe', in Barry, J., Baxter, B. and Dunphy, R. (eds) *Europe, Globalization and Sustainable Development*, pp. 94–112, London: Routledge

Benard, M. (1998) 'Electricity generation from renewable energy: the French experience', *Renewable Energy*, 15: 264–9

Bent, R., Bacher, A. and Thomas, I. (2002) 'Rules of the game', in Bent, R., Orr, L. and Baker, R. (eds) *Energy: Science, Policy and the Pursuit of Sustainability*, pp. 11–36, Washington, DC: Island Press

Bergek, A. and Jacobsson, S. (2003) 'The emergence of a growth industry: a comparative analysis of the German, Dutch and Swedish wind turbine industries', in Metcalfe, J. S. and Cantner, U. (eds) *Change, Transformation and Development*, pp. 197–227, Heidelberg: Physica Verlag

Birraux, C. and Le Déaut, J.-Y. (2001) *L'Etat actuel et les perspectives techniques des énergies renouvelables*, Paris: Office parlementaire d'évaluation des choix scientifiques et technologiques / Assemblée nationale, rapport no. 3415

Blázquez, G. G., Calero de Hoces, M. and Lehtinen, T. (2003) 'Policy networks of wind energy: the story of the first commercial wind farm in Spain', *Wind Engineering*, 27: 6, 461–72

Blühdorn, I. and Szarka, J. (2004) 'Managing strategic positioning choices: a reappraisal of the development paths of the French and German Green parties', *Journal of Contemporary European Studies*, 12: 3 (December), 303–19

BMU (2002) 'Strategy of the German government on the use of offshore wind energy in the context of its national sustainability strategy', Berlin: BMU

BMU (2004a) 'Environmental policy. Ecologically optimised extension of renewable energy utilisation in Germany – Summary', Berlin: BMU

BMU (2004b) 'Amending the Renewable Energy Sources Act (EEG). Key provisions of the new EEG on 21 July 2004', Berlin: BMU, http://www.bee-ev.de/uploads/eeg_begruendung_en.pdf Consulted 15.3.2006

BMU (2005) 'Environmental policy. Renewable energy sources in figures – national and international development. Status June 2005', http://erneuerbare-energien.de Consulted 5.4.2006

Børre Eriksen, P., Akhmatov, V. and Orths, A. (2006) 'Managing 23%: grid integration of wind power in Denmark', *Renewable Energy World*, 9: 4 (July–August), 214–27

Boston Consulting Group (2004) *Donner un nouveau souffle à l'éolien terrestre*, Paris: Boston Consulting Group

Breukers, S. (2006) *Changing Institutional Landscapes for Implementing Wind Power. A Geographical Comparison of Institutional Capacity Building: The*

Netherlands, England and North Rhine-Westphalia, Amsterdam: Amsterdam University Press

BTM Consult (2006) *International Wind Energy Development: World Market Update 2005*, Ringkøbing: BTM Consult ApS

Bustos, M. (2005) 'Spain's regulatory framework: helping wind reach 20,000 MW', *Wind Directions* (July/August), 21–3

BWE (2005) 'A clean issue – wind energy in Germany', Berlin: BWE

BWEA (2002) 'Wind can meet nuclear shortfall. Scope for 8% of UK electricity supply by 2010', http://www.bwea.com/media/news/energyreview.html Consulted 22.7.2005

BWEA (2004) 'Top myths about wind energy', http://www.bwea.com/energy/myths.html Consulted 14.12.2006

BWEA (2006) 'UKWED statistics', http://www.bwea.com/statistics/ Consulted 16.11.2006

Calero, R. and Carta, J. A. (2004) 'Action plan for wind energy development in the Canary Islands', *Energy Policy*, 32, 1185–97

Carbon Trust (2006) 'Policy frameworks for renewables. Analysis on policy frameworks to drive future investment in near and long-term renewable power in the UK', http://www.carbontrust.co.uk/Publications/CTC610.pdf Consulted 6.7.2006

Carruthers, D. (2005) 'From opposition to orthodoxy: the remaking of sustainable development', in Dryzek, J. and Schlosberg, D. (eds) *Debating The Earth: The Environmental Politics Reader*, pp. 285–300, Oxford: Oxford University Press

Chabot, B. (2001) 'Fair and efficient tariffs for wind energy: principles, method, proposal, data and potential consequences in France', in EWEA (eds) *Wind Energy for the New Millennium. Proceedings of the European Wind Energy Conference – Copenhagen, Denmark*, pp. 336–9, Brussels: EWEA

Chabot, B. (2003) 'Pourquoi et comment investir dans l'énergie éolienne en France', www.suivi-eolien Consulted 21.2.2005

Chabot, B. (2005) 'Comparaisons qualitatives et quantitatives des incitations au développement de l'électricité produite par sources d'énergies renouvelables', Valbonne: ADEME

Chabot, B. (2006) 'Assessment of fixed wind tariff system against tiered wind tariff systems', http://www.wind-works.org/FeedLaws/Chabot_Fixed_tariff_assesment2.pdf Consulted 12.6.2006

Chabot, B. and Buquet, L. (2006) 'Le développement de l'énergie éolienne en France en 2005', http://www.suivi-eolien.com/francais/DocsPDF/EolFrance 05V1.pdf Consulted 23.5.2006

Christensen, P. and Lund, H. (1998) 'Conflicting views of sustainability: the case of wind power and nature conservation in Denmark', *European Environment*, 8: 2, 1–6

CNE (2006) 'Informe mensual de ventas de energía del régimen especial en España, 30 junio 2006', http://www.cne.es/cne/ Consulted 4.7.2006

Cochet, Y. (2000) *Stratégies et moyens de développement de l'efficacité énergétique et des sources d'énergie renouvelables*, Paris: Documentation française

Collier, U. (1997) 'Windfall emissions reductions in the UK' in Collier, U. and Löfstedt, R. E. (eds) *Cases in Climate Change Policy. Political Reality in the European Union*, pp. 87–107, London: Earthscan

Country Guardian (2000a) 'Country Guardian's manifesto', http://www.countryguardian.net/Manifesto.htm Consulted 11.12.2006

Country Guardian (2000b) 'The case against wind farms' http://www.countryguardian.net/case.htm Consulted 18.2.2004

Country Guardian (2003) 'The windfarm white elephant', http://www.countryguardian.net/The%20windfarm%20white%20elephant.htm Consulted 18.2.2004

Country Guardian (2004) 'Windfarm wars: how to fight a windfarm', http://www.countryguardian.net/How%20to%20fight%20a%20windfarm.htm Consulted 17.6.2004

Country Guardian (2006) 'Home page', http://www.countryguardian.net/cg.htm Consulted 7.12.2006

Country Guardian / Etherington, J. R. (2006) 'The case against wind farms – 2006 edition', http://www.countryguardian.net/Case%202006.htm Consulted 7.12.2006

Countryside Agency (2003) 'Objection by the Countryside Agency to the proposed Whinash wind farm', http://www.persona.uk.com/whinash/CORE_DOCS/CD_221_CA.pdf Consulted 9.5.2005

Countryside Commission (1991) 'Wind energy development and the landscape', Manchester: Countryside Commission

CPRE (2006) 'Policy position statement: onshore wind turbines', http://www.cpre.org.uk Consulted 14.12.2006

CPRW (2005a) 'Renewable energy: offshore wind – current CPRW policy', http://www.cprw.org.uk/renewenergy_offshore.htm Consulted 14.12.2006

CPRW (2005b) 'Welsh Affairs Select Committee: inquiry into energy in Wales. Representation on behalf of the CPRW', http://www.cprw.org.uk/renewenergy.htm Consulted 14.12.2006

Cullingworth, B. and Nadin, V. (2002) *Town and Country Planning in the UK*, London: Routledge

Dalton, R. J. (1993) 'The environmental movement in Western Europe', in Kamieniecki, S. (ed.) *Environmental Politics in the International Arena. Movements, Parties, Organisations and Policy*, pp. 41–68, New York: State University of New York

Danielsen, O. (1995) 'Large-scale wind power in Denmark', *Land Use Policy*, 12: 1, 60–2

Danish Energy Agency (1999) 'Wind power in Denmark: technology, policies and results', Copenhagen: Ministry of Environment and Energy, http://www.ens.dk/graphics/Publikationer/Forsyning_UK/wind-power99.pdf Consulted 15.10.2004

Danish Energy Agency (2001) 'The green certificate market in Denmark: status of implementation', Copenhagen: Danish Energy Agency

Danish Energy Authority (2002) 'Wind energy in Denmark. Status 2001', Copenhagen: Ministry of Environment and Energy

Danish Energy Authority (2005) 'Offshore wind power: Danish experience and solutions', http://www.ens.dk/graphics/Publikationer/Havvindmoeller/uk_vindmoeller_okt05/pdf/havvindmoellerapp_GB-udg.pdf Consulted 18.8.2006

Danish Ministry of Energy (1990) *Energy 2000. A Plan of Action for Sustainable Development*, Copenhagen: Danish Energy Agency

Danish Ministry of Environment and Energy (1996) *Energy 21. The Danish Government's Action Plan for Energy 1996*, Copenhagen: Danish Ministry of Environment and Energy

DeCarolis, J. F. and Keith, D. W. (2005) 'The costs of wind's variability: is there a threshold?', *The Electricity Journal*, 18: 1 (January–February), 69–77

DENA (2005) 'Planning of the grid integration of wind energy in Germany onshore and offshore up to the year 2020 (dena grid study): summary of the essential results of the study', http://www.deutsche-energie-agentur.de/page/fileadmin/DeNA/dokumente/Programme/Kraftwerke_Netze/dena_Grid_Study_Summary_2005-03-23.pdf Consulted 24.4.2006

DENA (2006) *Renewables Made in Germany*, Berlin: DENA

Department of the Environment (1993) 'PPG22 – Planning Policy Guidance Note: Renewable Energy' http://www.odpm.gov.uk/stellent/groups/odpm_planning/documents/page/odpm_plan_606910.pdf Consulted 3.12.2004

DEWI (2006) 'Wind energy use in Germany – status 31.12.2005', http://www.dewi.de/ Consulted 10.5.2006

DGEMP-DIDEME (2004) 'L'éolien en France: une montée en puissance', http://www.industrie.gouv.fr/energie/renou/eolien-enquete04.htm Consulted 21.2.2005

Dickson, D. (1974) *Alternative Technology and the Politics of Technical Change*, London: Fontana

Dinica, V. (2002) 'Spain', in Reiche, D. (ed.) *Handbook of Renewable Energies in the EU. Case Studies of All Member States*, pp. 211–26, Frankfurt: Peter Lang

Dobson, A. (1998) *Justice and the Environment. Conceptions of Environmental Sustainability and Dimensions of Social Justice*, Oxford: OUP

Dodd, J. (2006) 'Radar exclusion zones: France put out of bounds', *Windpower Monthly*, 22: 4 (April), 46

DTI (2004) *Renewable Supply Chain Gap Analysis: Summary Report*, London: DTI

DTI (2006a) 'Renewables obligation order 2006 – final decisions', http://www.dti.gov.uk/renewables/policy_pdfs/roo2006finalpositionpaper.pdf Consulted 24.4.2006

DTI (2006b) 'Wind power: 10 myths explained', http://www.dti.gov.uk/energy/sources/renewables/renewables-explained/intro/intro-faqs/Wind%20power%2010%20myths%20explained/page16060.html Consulted 14.12.2006

Dunion, K. (2003) *Troublemakers: The Struggle for Environmental Justice in Scotland*, Edinburgh: Edinburgh University Press

DWIA (2002) 'DWIA Annual Report', *Windpower Note*, 27 (March)

DWIA (2004) ' Merger for the future – interview with CEO Svend Sigaard, Vestas Wind Systems A/S', http://www.windpower.org/en/news/interview040415.htm Consulted 12.8.2004

DWIA (2005) 'Annual Report of the DWIA', http://www.windpower.org/media(775,1033)/annual_report_2004.pdf Consulted 18.8.2006

DWIA / Holst, J. L. (2006) 'Denmark – wind power hub', Paper given at the Hamburg Wind Fair, May 2006

E.ON Netz (2004) 'Wind report 2004', http://www.eon-netz.com Consulted 21.3.2005

E.ON Netz (2005) 'Wind report 2005', http://www.eon-netz.com/Ressources/downloads/EON_Netz_Windreport2005_eng.pdf Consulted 26.1.2007

ECJ (2001) 'Judgement of the Court, 13 March 2001 in case C-379/98 PreussenElektra AG and Schleswag AG', http://curia.europa.eu/jurisp/cgi-bin/form.pl?lang=en&Submit=Submit&alldocs=alldocs&docj=docj&docop=docop&docor=docor&docjo=docjo&numaff=C-379%2F98&datefs=&datefe=&nomusuel=&domaine=&mots=&resmax=100 Consulted 5.10.2006

Ecofys (2002) *Green Energy in Europe – Strategic Prospects to 2010*, London: Reuters Business Insight

Edwards, P. D. (1994) 'The UK's first windfarm – the birth of an industry', *Renewable Energy*, 5: 1, 637–41

EEA (2004) 'Greenhouse gas emission trends and projections in Europe 2004', Copenhagen: EEA report no. 5, http://reports.eea.eu.int/eea_report_2004_5/en/GHG_emissions_and_trends_2004.pdf Consulted 12.1.2005

Elliott, D. (1992) 'Renewables and the privatisation of the UK electricity supply industry', *Energy Policy*, 20: 3 (March), 257–68

Elliott, D. (2003) *Energy, Society and Environment*, London: Routledge

Elliott, D. (2005) 'Comparing support for renewable power', in Lauber, V. (ed.) *Switching to Renewable Power: A Framework for the 21st Century*, pp. 219–27, London: Earthscan

Ellis, G. (2004) 'Discourses of objection: towards an understanding of third-party rights in planning', *Environment and Planning A*, 36, 1549–70

English Nature, RSPB, WWF-UK and BWEA (2001) 'Wind farm development and nature conservation. A guidance document for nature conservation organisations and developers when consulting over wind farm proposals in England', Godalming, Surrey: WWF-UK in association with English Nature, RSPB and BWEA

Enzensberger, N., Wietschel, and Rentz, O. (2002) 'Policy instruments fostering wind energy projects – a multi-perspective evaluation approach', *Energy Policy*, 30: 9 (July), 793–801

Eriksen, P. B. and Hilger, C. (2005) 'Wind power in the Danish power system', in Ackermann, T. (ed.) *Wind Power in Power Systems*, pp. 199–232, Chichester: John Wiley

Eurelectric (2004) 'A quantitative assessment of direct support schemes for renewables', http://www.eurelectric.org Consulted 21.3.2005

Europa Nostra (2004a) 'Wind turbines: careless planning causes serious damage to the environment', http://www.europanostra.org/lang_en/index.html Consulted 14.12.2006

Europa Nostra (2004b) 'Declaration on the impact of wind power on the countryside', http://www.europanostra.org/lang_en/index.html Consulted 8.1.2007

European Commission (1997) 'Energy for the future: renewable sources of energy. White paper for a Community strategy and action plan', COM (97) 599 final http://ec.europa.eu/energy/library/599fi_en.pdf Consulted 15.1.2007

European Commission (2005) 'The support of electricity from renewable energy sources', COM (2005) 627 final http://ec.europa.eu/energy/res/biomass_action_plan/doc/2005_12_07_comm_biomass_electricity_en.pdf Consulted 12.5.2006

Everett, B. and Boyle, G. (2004) 'Integration', in Boyle, G. (ed.) *Renewable Energy: Power For A Sustainable Future*, pp. 384–432, Oxford: OUP

EWEA (2004) 'On the future of EU support systems for the promotion of electricity from renewable energy sources', Brussels: EWEA

EWEA (2006) 'Special No Fuel Edition', *Wind Directions* (January–February)

EWEA and Greenpeace (2002) 'Wind force 12: A blueprint to achieve 12% of the world's electricity from wind power by 2020', Brussels: EWEA and Greenpeace

Farstad, H. and Ward, J. (1984) 'Wind energy in Denmark', in Baumgartner, T. and Burns, T. R. (eds) *Transitions to Alternative Energy Systems*, pp. 91–124, Boulder: Westview Press

Faulin, J. *et al.* (2006) 'The outlook for renewable energy in Navarre: an economic profile', *Energy Policy*, 34, 2201–16

Flam, H. (ed.) (1994) *States and Anti-nuclear Movements*, Edinburgh: Edinburgh University Press

FoE (2003) 'Goodbye nuclear, hello wind', http://www.foe.co.uk/resource/press_releases/goodbye_nuclear_hello_wind.html Consulted 6.9.2004

FoE (2004) 'Why nuclear power is not an achievable and safe answer to climate change', London: FoE

Frandsen, S. and Andensen, P. D. (1996) 'Wind farm progress in Denmark', *Renewable Energy*, 9, 848–52

García-Cebrián, L. I. (2002) 'The influence of subsidies on the production process: the case of wind energy in Spain', *The Electricity Journal* (May), 79–86

Garrigues, B. (2002) '700MW pour la Navarre', *Systèmes solaires*, 149, 74–7

German Parliament (2000) 'Act on granting priority to renewable energy sources. Renewable Energy Sources Act', Berlin: German Parliament

Gipe, P. (1995) *Wind Energy Comes of Age*, New York: John Wiley

Grotz, C. (2002) 'Germany', in Reiche, D. (ed.) *Handbook of Renewable Energies in the EU. Case Studies of All Member States*, pp. 107–23, Frankfurt: Peter Lang

GURELUR (2006) 'Windpower is fine, but not at any price', http://www.gurelur.org/wind%20power.htm, Consulted 8.1.2007

GWEC (2006) *Global Wind 2005 Report*, http://www.gwec.net/fileadmin/documents/Publications/GWEC-Global_Wind_05_Report_low_res_01.pdf Consulted 9.9.2006

GWEC and Greenpeace (2005) 'Windforce 12. A Blueprint to achieve 12% of the world's electricity from wind power by 2020. Report 2005', http://www.greenpeace.org/raw/content/international/press/reports/windforce-12-2005.pdf Consulted 15.5.2006

Haas, R. *et al.* (2004) 'How to promote renewable energy systems successfully and effectively', *Energy Policy*, 32, 833–9

Hajer, M. A. (1995) *The Politics of Environmental Discourse: Ecological Modernization and the Policy Process*, Oxford: Clarendon Press

Hajer, M. A. (2005) 'Coalitions, practices and meaning in environmental politics', in Howarth, D. R. and Torfing, J. (eds) *Discourse Theory in European Politics. Identity, Policy and Governance*, pp. 297–315, London: Palgrave

Halkema, J. A. (2006) 'Wind energy: facts and fiction. A half truth is a whole lie', http://www.countryguardian.net/halkema-windenergyfactfiction.pdf Consulted 7.12.2006

Haugland, T., Bergesen, H. O. and Roland, K. (1998) *Energy Structures and Environmental Futures*, Oxford: OUP

Héritier, A., Knill, C. and Mingers, S. (1996) *Ringing the Changes in Europe. Regulatory Competition and Redefinition of the State: Britain, France, Germany*, Berlin: Walter de Gruyter

Heymann, M. (1999) 'A fight of systems? Wind power and electric power systems in Denmark, Germany and the USA', *Centaurus*, 41: 1–2, 112–36

Hinshelwood, E. *et al.* (2000) 'Community funded wind power – the missing link in UK wind farm development?', *Wind Engineering*, 24: 4, 299–305

Holttinen, H. and Tuhkanen, S. (2004) 'The effect of wind power on CO_2 abatement in the Nordic countries', *Energy Policy*, 32, 1639–52

Hopkins, W. (1999) 'Small to medium sized wind turbines: local use of a local resource', *Renewable Energy*, 16, 944–7

Hoppe-Kilpper, M. and Steinhäuser, U. (2002) 'Wind landscapes in the German milieu', in Pasqualetti, M. J., Gipe, P., Righter, R. W. (eds) *Wind Power in View: Energy Landscapes in a Crowded World*, pp. 83–99, London: Academic Press

House of Commons Environmental Audit Committee (2006) 'Keeping the lights on: nuclear, renewables and climate change', http://www.publications.parliament.uk/pa/cm200506/cmselect/cmenvaud/584/584i.pdf Consulted 3.5.2006

Howlett, M. (2002) 'Do networks matter? Linking policy network structure to policy outcomes', *Canadian Journal of Political Science*, 35: 2 (June), 235–67

Hurtado, J. P. *et al.* (2004) 'Spanish method of visual impact evaluation in wind farms', *Renewable and Sustainable Energy Reviews*, 8, 483–91

Hvelplund, F. (2001a) *Renewable Energy Governance Systems*, Aalborg: Aalborg University

Hvelplund, F. (2001b) *Electricity Reforms, Innovative Democracy and Technological Change*, Aalborg: Aalborg University

Hvelplund, F. (2001c) 'Political prices or political quantities? A comparison of renewable energy support systems', *New Energy*, 5, 18–23

Hvelplund, F. (2002) 'Denmark' in Reiche, D. (ed.) *Handbook of Renewable Energies in the EU. Case Studies of All Member States*, pp. 63–75, Frankfurt: Peter Lang

Hvelplund, F. (2005) 'Renewable energy: political prices or political quantities', in Lauber, V. (ed.) *Switching to Renewable Power: A Framework for the 21st Century*, pp. 228–45, London: Earthscan

Hvidtfelt Nielsen, K. (2005) 'Danish wind power policies from 1976 to 2004: a survey of policy making and techno-economic innovation', in Lauber, V. (ed.) *Switching to Renewable Power: A Framework for the 21st Century*, pp. 99–121, London: Earthscan

Ibenholt, K. (2002) 'Explaining learning curves for wind power', *Energy Policy*, 30, 1181–9

IDAE (1999) *Plan de fomento de la energías renovables en España*, Madrid: IDAE

IDAE (2005a) 'The Spanish Renewable Energy Plan 2005–2010 – summary', http://www.idae.es/central.asp?a=p3&i=en# Consulted 6.6.2006

IDAE (2005b) *Boletín IDAE: Eficiencía Energética y Energías Renovables*, no. 7 (septiembre), Madrid: IDAE

IDAE (2006) 'Wind energy in Spain 2005: current status and prospects', Madrid: IDAE

IEA (2002) *IEA Wind Energy Annual Report 2001*, Paris: OECD/IEA http://www.ieawind.org Consulted 17.10.2003

IEA (2005a) *IEA Wind Energy Annual Report 2004*, Paris: OECD/IEA, http://www.ieawind.org Consulted 18.7.2005

IEA (2005b) *Electricity Information*, Paris: OECD/IEA

IEA (2006a) *IEA Wind Energy Annual Report 2005,* Paris: OECD/IEA, http:// www.ieawind.org Consulted 18.8.2006

IEA (2006b) *Energy Policies of IEA countries: Denmark 2006 Review,* Paris: OECD/IEA

Infield, D. (1995) 'Wind power – a major resource for the UK', *Power Engineering,* 9: 4 (August), 181–7

Jaccard, M. (2005) *Sustainable Fossil Fuels. The Unusual Suspect in the Quest for Clean and Enduring Energy,* Cambridge: CUP

Jackson, J. (2003) 'A cry for better forecasters in Denmark', *Windpower Monthly* (December), 40–2

Jianxiang, Y. (2006) 'Flying the wind flag high in Beijing', *Windpower Monthly,* 22: 9 (September), 70–4

Jianxiang, Y. and Knight, S. (2006) 'Chinese wind turbine suppliers sweep the board', *Windpower Monthly,* 22: 11 (November), 27–8

Johnson, D. (2004) 'Ecological modernization, globalization and Europeanization', in Barry, J., Baxter, B. and Dunphy, R. (eds) *Europe, Globalization and Sustainable Development,* pp. 152–67, London: Routledge

Jørgensen, U. and Karnøe, P. (1995) 'The Danish wind-turbine story: technical solutions to political visions', in Rip, A. R., Misa, T. J. and Schot, J. (eds) *Managing Technology in Society. The Approach of Constructive Technology Assessment,* pp. 57–82, London: Pinter

Kamp, L., Smits, R. E. H. M. and Andriesse, C. D. (2004) 'Notions on learning applied to wind turbine development in the Netherlands and Denmark', *Energy Policy,* 32, 1625–37

Karnøe, P. (1990) 'Technological innovation and industrial organisation in the Danish wind industry', *Entrepreneurship and Regional Development,* 2, 105–23

Kirkby, J., O'Keefe, P. and Timberlake, L. (eds) (1995) *The Earthscan Reader in Sustainable Development,* London: Earthscan

Kitschelt, H. (1986) 'Political opportunity structure and political protest: antinuclear movements in four democracies', *British Journal of Political Science,* 16: 1, 57–83

Knight, S. (2006) 'Major utility taken to task by energy regulator', *Windpower Monthly,* 22: 10 (October), 29–30

Krewitt, W. (2002) 'External costs of energy – do the answers match the questions? Looking back at ten years of ExternE', *Energy Policy,* 30: 10 (August), 839–48

Krohn, S. (1998) 'Creating a local wind industry. Experience from four European countries', http://www.windpower.org Consulted 27.7.2003

Laali, A. R. and Benard, M. (1999) 'French wind power generation programme EOLE 2005: results of the first call for tenders', *Renewable Energy,* 16, 805–10

Lake, R. W. (1993) 'Rethinking NIMBY', *Journal of the American Planning Association,* 59: 1 (Winter), 87–96

Langhelle, O. (2000) 'Why ecological modernisation and sustainable development should not be conflated', *Journal of Environmental Policy and Planning,* 2, 303–22

Larcher, G. and Revol, H. (2003) 'Energie: quelle politique française pour la prochaine législature?', Paris: Les Rapports du Sénat, no. 79

Lauber, V. (2002) 'The different concepts of promoting RES-electricity and their political careers', in Biermann, F., Brohm, R. and Dingwerth, K. (eds)

Proceedings of the 2001 Berlin Conference on the Human Dimensions of Global Environmental Change: Global Environmental Change and the Nation State, pp. 296–304, Potsdam: Potsdam Institute for Climate Impact Research

Lauber, V. (2004) 'REFIT and RPS: options for a harmonised Community framework', *Energy Policy,* 32, 1405–14

Lauber, V. (2005a) 'European Union policy towards renewable power', in Lauber, V. (ed.) *Switching to Renewable Power: A Framework for the 21st Century,* pp. 203–16, London: Earthscan

Lauber, V. (2005b) 'Tradeable certificate systems and feed-in tariffs: expectation versus performance', in Lauber, V. (ed.) *Switching to Renewable Power: A Framework for the 21st Century,* pp. 246–63, London: Earthscan

Lee, R. (2002) 'Environmental impacts of energy use', in Bent, R., Orr, L. and Baker, R. (eds) *Energy: Science, Policy and the Pursuit of Sustainability,* pp. 77–108, Washington, DC: Island Press

Lehmann, K. P. (2003) 'Strategic options for the wind energy market', *Renewable Energy World,* 6: 3, (May–June), 38–49

Lindsay, R. (2005) 'Lewis Wind Farm proposals: observations on the official Environmental Impact Statement. A report commissioned by the RSPB', University of East London http://www.rspb.org.uk/Images/lewiswindfarm-peatland_tcm5-69783.pdf Consulted 25.7.2005

Little, K. (ed.) (2003) *The Battle for Cefn Croes. How New Labour's Energy Policy Conspired to Destroy the Landscape of the Welsh Cambrian Mountains,* Tan-y-Glog: Cefn Croes Publications, http:www/.users.globalnet.co.uk/~hills/cc/book/index.htm Consulted 24.3.2004

Lovins, A. B. (1977) *Soft Energy Paths: Towards a Durable Peace,* Cambridge, Mass.: Ballinger Publishing Company

Luckin, B. (1990) *Questions of Power. Electricity and Environment in Inter-war Britain,* Manchester: Manchester University Press

Luther, M., Radtke, U. and Winter, W. R. (2005) 'Wind power in the German power system: current status and future challenges in maintaining quality of supply', in Ackermann, T. (ed.) *Wind Power in Power Systems,* pp. 233–55, Chichester: John Wiley

MacKerron, G. (2003) 'Electricity in England and Wales: efficiency and equity', in Glachant, J. M. and Finon, D. (eds) *Competition in European Electricity Markets: A Cross-Country Comparison,* pp. 41–56, Cheltenham: Edward Elgar

Majone, G. (1989) *Evidence, Argument and Persuasion in the Policy Process,* New Haven: Yale University Press

Man, R. de (1987) 'United Kingdom energy policy and forecasting: technocratic conflict resolution', in Baumgartner, T. and Midttun, A. (eds) *The Politics of Energy Forecasting: A Comparative Study of Energy Forecasting in Western Europe and North America,* pp. 110–34, Oxford: Oxford University Press

Mann, E. (2003) *Force 10: Political Will versus Landscape Protection,* http://www.wind-farm.co.uk/force10.pdf Consulted 8.1.2007

Massy, J. (2005) 'Optimism tempered by new reality in Britain', *Windpower Monthly,* 21: 12 (December), 52–6

Massy, J. (2006) 'Flight trials test solution to radar interference', *Windpower Monthly,* 22: 7 (July), 25–6

May, H. (2006) 'The world goes shopping', *New Energy*, 2, 46–50

Meadowcroft, J. (2000) 'Sustainable development: a new(ish) idea for a new century?', *Political Studies*, 48, 370–87

Meadows, D. H., Meadows, D. L., Randers, J. and Behrens, W. W. (1972) *The Limits to Growth. A Report for the Club of Rome's Project on the Predicament of Mankind*, London: Pan

Menanteau, P. (2000) 'L'énergie éolienne: la réussite d'une dynamique d'innovations incrémentales', in Bourgeois, B. *et al. Energie et changement technologique. Une approche évolutionniste*, pp. 221–47, Paris: Economica

Meyer, N. I. (2003) 'European schemes for promoting renewables in liberalised markets', *Energy Policy*, 31: 7, 665–76

Meyer, N. L. and Koefoed, A. L. (2003) 'Danish energy reform: policy implications for renewables', *Energy Policy*, 31: 7 (June), 597–607

Mez, L. (1995) 'Reduction of exhaust gases at large combustion plants in the Federal Republic of Germany', in Jänicke, M. and Weidner, H. (eds) *Successful Environmental Policy: A Critical Evaluation of 24 Cases*, pp. 173–86, Berlin: Edition Sigma

Mez, L. (1997) 'The German electricity reform attempts: reforming co-optive networks', in Midttun, A. (ed.) *European Electricity Systems in Transition. A Comparative Analysis of Policy and Regulation in Europe*, pp. 231–52, Oxford: Elsevier

Mez, L. and Midttun, A. (1997) 'The politics of electricity regulation', in Midttun, A. (ed.) *European Electricity Systems in Transition. A Comparative Analysis of Policy and Regulation in Europe*, pp. 307–31, Oxford: Elsevier

Mez, L. and Piening, A. (2002) 'Phasing-out nuclear power generation in Germany: policies, actors, issues and non-issues', *Energy and Environment*, 13: 2, 161–81

Mez, L., Midttun, A. and Thomas, S. (1997) 'Restructuring electricity systems in transition', in Midttun, A. (ed.) *European Electricity Systems in Transition. A Comparative Analysis of Policy and Regulation in Europe*, pp. 3–12, Oxford: Elsevier

Michaelowa, A. (2005) 'The German wind lobby: how to promote costly technological change successfully', *European Environment*, 15, 192–9

Midttun, A. (1997) 'Regulation paradigms and regulation practice', in Midttun, A. (ed.) *European Electricity Systems in Transition. A Comparative Analysis of Policy and Regulation in Europe*, pp. 279–305, Oxford: Elsevier

Midttun, A. and Koefoed, A. L. (2003) 'Greening of electricity in Europe: challenges and developments', *Energy Policy*, 31: 7, 677–87

Milborrow (2007) 'Back to being a model of stability', *Windpower Montly*, 23: 1 (January), 47–50

Milborrow, D. and Harrison, L. (2004) 'The real cost of integrating wind', *Windpower Monthly* (February), 35–9

Milborrow, D. (2004) 'The real costs of wind versus nuclear', *Windstats Newsletter*, 17: 2 (Spring), 1–9

Ministère de l'Economie, des Finances et de l'Industrie (2005a) 'Eolien terrestre: Francis Loos annonce les résultats du premier appel d'offre: sept projets représentent une puissance cumulée de 278,35',http://www.industrie.gouv.fr/ energie/renou/ textes/com-eolien-terre-long.htm Consulted 16.6.2006

Ministère de l'Economie, des Finances et de l'Industrie (2005b) 'L'éolien en France: une montée en puissance confirmée', http://www.industrie.gouv.fr/ energie/ renou/eolien-enquete.htm Consulted 16.6.2006

Mitchell, C. (2000) 'The England and Wales non-fossil fuel obligation: history and lessons', *Annual Review of Energy and the Environment*, 25, 285–312

Mitchell, C., Bauknecht, D. and Connor, P. M. (2006) 'Effectiveness through risk reduction: a comparison of the Renewable Obligation in England and Wales and the feed-in system in Germany', *Energy Policy*, 34, 297–305

Mitchell, C. and Connor, P. (2004) 'Renewable energy policy in the UK 1990–2003', *Energy Policy*, 32, 1935–47

National Audit Office (2005) *Renewable Energy. Report by the Comptroller and Auditor General*, http://www.nao.org.uk/publications/nao_reports/04-05/ 0405210.pdf Consulted 14.3 2005

NFPA (2006) 'Home page', http://www.nfpa.co.uk/index.cfm Consulted 25.4.2005

Nielsen, F. B. (2002) 'A formula for success in Denmark', in Pasqualetti, M. J., Gipe, R., W. Righter, R. W. (eds) *Wind Power in View: Energy Landscapes in a Crowded World*, pp. 115–32, London: Academic Press

Norris, P. (1997) 'Are we all green now? Public opinion on environmentalism in Britain', *Government and Opposition*, 32: 3 (Summer), 320–39

O'Riordan, T. and Voisey, H. (eds) (1998) *The Transition to Sustainability. The Politics of Agenda 21 in Europe*, London: Earthscan

ODPM (2004a) *Planning Policy Statement 22: Renewable Energy*, London: The Stationery Office, http://www.communities.gov.uk/pub/910/PlanningPolicy Statement22RenewableEnergy_id1143910.pdf Consulted 3.5.2005

ODPM (2004b) *Planning for Renewable Energy. A Companion Guide to PPS22*, London: The Stationery Office, http://www.communities.gov.uk/pub/915/ PlanningforRenewableEnergyACompanionGuidetoPPS22_id1143915.pdf Consulted 3.5.2005

OECD (2002) *Environmental Performance Reviews: United Kingdom*, Paris: OECD

OFGEM (2006a) 'Renewables obligation: third annual report', http://www. ofgem.gov.uk/temp/ofgem/cache/cmsattach/15383_ROannualreport.pdf Consulted 24.4.2006

OFGEM (2006b) 'The Renewables Obligation 2004–5: facts and figures', http://www.ofgem.gov.uk/temp/ofgem/cache/cmsattach/14028_56.pdf Consulted 24.4.2006

OFGEM (2007) 'Reform of the Renewables Obligation 2006: Ofgem's response', http://www.ofgem.gov.uk/temp/ofgem/cache/cmsattach/18363_ROrespJan.pd f Consulted 1.2.2007

Olesen, G., Maegaard, P. and Kruse, J. (2003) 'Wind energy and local financing: the Danish experience', Hurup Thy: Folkecentre for Renewable Energy

OSS (2006) 'Wind turbines on common land', http://www.oss.org.uk/publica-tions/infosht/a4.htm Consulted 14.12.2006

Oxera (2005) 'Renewables support policies in selected countries', http://www. nao.org.uk/publications/nao_reports/04-05/0405210_selected_ countries_renewables.pdf Consulted 14.3 2005

Poppe, M. and Cauret, L. (1997) 'The French electricity regime', in Midttun, A. (ed.) *European Electricity Systems in Transition. A Comparative Analysis of Policy and Regulation in Europe*, pp. 199–229, Oxford: Elsevier

RA (2005) 'A renewable energy policy for the RA (England)', http://www.ramblers.org.uk/countryside/energy/Energypolicy05.pdf Consulted 14.12.2006

RA (2006) 'Ramblers' Association Scotland: energy policy statement', http://www.ramblers.org.uk/countryside/energy/RAS_REPolicyJune06.pdf Consulted 14.12.2006

Rand, M. and Clarke, A. (1990) 'The environmental and community impacts of wind energy in the UK', *Wind Engineering*, 14: 5, 319–30

Redclift, M. (1987) *Sustainable Development: Exploring the Contradictions*, London: Routledge

Redlinger, R. Y., Dannemand, A. P. and Morthorst, P. E. (2002) *Wind Energy in the 21st Century. Economics, Policy, Technology, and the Changing Electricity Industry*, Basingstoke: Palgrave

REF (2004a) 'Renewable Energy Foundation', http://www.ref.org.uk/ Consulted 7.12.2006

REF (2004b) '2005–2006 Review of the Renewables Obligation', http://www.ref.org.uk/images/pdfs/REF.ROC.pdf Consulted 14.3.2005

REF/ Hall, M. J. (2006) 'A guide to calculating the carbon dioxide debt and payback time for wind farms', http://www.viewsofscotland.org/oban_demo/PeatAudit%20-%20Guide.pdf Consulted 7.12.2006

Reiche, D. and Bechberger, M. (2004) 'Policy differences in the promotion of renewable energies in the EU', *Energy Policy*, 32, 843–9

Rein, M. and Schön, D. (1993) 'Reframing policy discourse', in Fischer, F. and Forester, J. (eds) *The Argumentative Turn in Policy Analysis and Planning*, pp. 145–66, Durham: Duke University Press

République française (2003) 'Arrêté du 7 mars 2003 relatif à la programmation pluriannuelle des investissements de production d'électricité', *Journal Officiel*, 65 (18 March), 4692, http://www.francaise-d-eoliennes.com/docs/arrete20030307.pdf Consulted 16.6.2006

Rickerson, W. (2002) 'German renewable energy feed-in tariffs policy overview', http://www.ontario-sea.org/ARTs/Germany/GermanyRickerson.html Consulted 10.11.2004

Ringel, M. (2006) 'Fostering the use of renewable energies in the European Union: the race between feed-in tariffs and green certificates', *Renewable Energy*, 31, 1–17

Río, P. del and Gual, M. A. (2006) 'An integrated assessment of the feed-in tariff system in Spain', *Energy Policy*, in press

Risø National Laboratory (1989) *European Wind Atlas*, Brussels: Commission of the EC

Rose, D. M. H. (2006) 'Report to the Secretaries of State for Trade and Industry; and for Environment, Food and Rural Affairs: Whinash wind farm', Bristol: The Planning Inspectorate, http://www.persona.uk.com/whinash/DECISION/Whinash_Inspectors_Report.pdf Consulted 15.3.2006

Rowlands, I. H. (2005) 'The European directive on renewable electricity: conflicts and compromises', *Energy Policy*, forthcoming

RSPB (2003) 'Successful offshore wind farms bids raise serious concerns for birds', http://www.rspb.org.uk/policy/windfarms.roundtwo.asp Consulted 18.7.2005

RSPB (2004) 'Information – wind farms and birds', http://www.rspb.org.uk/Images/Pages%20from%20Wind%20farms%20and%20birds_tcm5-51248.pdf Consulted 18.7.2005

RSPB (2005) 'RSPB objection to original Lewis wind power proposal', http://www.rspb.org.uk/Images/lewis_tcm5-67068.pdf Consulted 8.1.2007

RSPB and BirdLife International (2003) 'Windfarms and birds: an analysis of the effects of windfarms on birds, and guidance on environmental assessment criteria and site selection issues', Strasbourg: Council of Europe, T-PVS/Inf (2003) 12

Sabatier, P. A. (1993) 'Policy change over a decade or more', in Sabatier, P. and Jenkins-Smith, H. C. (eds) *Policy Change and Learning. An Advocacy Coalition Approach*, pp. 13–39, Boulder: Westview Press

Sabatier, P. A. and Jenkins-Smith, H. C. (1993) 'The advocacy coalition framework: assessment, revisions and implications for scholars and practitioners' in Sabatier, P. and Jenkins-Smith, H. C. (eds) *Policy Change and Learning. An Advocacy Coalition Approach*, pp. 211–34, Boulder: Westview Press

Sabatier, P. A. and Jenkins-Smith, H. C. (1999) 'The advocacy coalition framework: an assessment', in Sabatier, P. (ed.) *Theories of the Policy Process*, pp. 117–66, Boulder: Westview Press

Sawin, J. (2004) 'National policy instruments. Policy lessons for the advancement and diffusion of renewable energy technologies around the world', International Conference for Renewable Energies, Bonn, http://www.renewables2004.de, Consulted 10.11.04

Scottish Executive (2000) 'National Planning Policy Guideline – NPPG6 Renewable Energy Developments', http://www.scotland.gov.uk/library3/planning/nppg/nppg6.pdf Consulted 3.12.2004

Secretary of State for Trade and Industry (2003) *Our Energy Future – Creating A Low Carbon Economy*, London: HMSO

SEO/BirdLife (2005) 'Eólicas sí, pero no en áreas sensibles para las aves', http://www.seo.org/media/docs/Posici%C3%B3n%20e%C3%B3licas-aves.PDF Consulted 8.1.2007

SEO/BirdLife (2006a) 'SEO/BirdLife solicita que no se autoricen parques eólicos en áreas importantes para las aves en la Comunidad Valenciana', http://www.seo.org/programa_noticia_ficha.cfm?idPrograma=21&idArticulo=1197&CFID=41745&CFTOKEN=66649622&jsessionid=aa3057f5441936456b1b Consulted 8.1.2007

SEO/BirdLife (2006b) 'SEO/BirdLife y ADENEX consideran improcedente la forma en que la Junta de Extremadura pretende autorizar la explotación de energía eólica en la región', http://www.seo.org/sala_detalle.cfm?idSala=1588&CFID=41745&CFTOKEN=66649622&jsessionid=aa3057f5441936456b1b Consulted 8.1.2007

Serrallés, R. J. (2005) 'Electric energy restructuring in the European Union: integration, subsidiarity and the challenge of harmonisation', *Energy Policy*, in press

Sesto, E. and Lipman, N. H. (1992) 'Wind energy in Europe', *Wind Engineering*, 16: 1, 35–47

Sharman, H. (2005) 'Why wind power works for Denmark', *Civil Engineering*, 158 (May), 66–72

Simpson, D. (2004) 'Tilting at windmills: the economics of wind power', The David Hume Institute, Hume Occasional Paper no. 65, http://www.davidhumeinstitute.com/DHI%20Website/publications/hop/Wind%20Power%20paper.pdf Consulted 10.5.2004

Sjödin, J. and Grönkvist, S. (2004) 'Emissions accounting for use and supply of electricity in the Nordic market', *Energy Policy*, 32, 1555–64

Smith, A. (2000) 'Policy networks and advocacy coalitions: explaining change and stability in UK industrial pollution policy?', *Environment and Planning C: Government and Policy*, 18, 95–114

SNH (2002) 'Statement: SNH's policy on renewable energy', Perth: SNH

SNH (2003) 'Guidance: cumulative effect of windfarms', Perth: SNH

SNH (2004) 'Renewables trends in Scotland: statistics and analysis, http://www.snh.org.uk/pdfs/strategy/renewable/sr-rt.pdf Consulted 29.6.2005

SNH (2005a) 'Policy statement: strategic locational guidance for onshore wind farms in respect of natural heritage', Perth: SNH

SNH (2005b) 'RSE inquiry into energy issues for Scotland', Edinburgh: SNH

Sørensen, H.-C. and Hansen, L. K. (2001) 'Concerted action on offshore wind energy in Europe: Social acceptance, environmental impact and politics', http://www.emu-consult.dk/includes/c5report.pdf Consulted 21.9.2003

Spaargaren, G. and Mol, A. P. J. (1992) 'Sociology, environment and modernity: ecological modernisation as a theory of social change', *Society and Natural Resources*, 5, 323–44

SPPEF (2002) 'Manifeste du groupe de réflexion et de propositions sur l'éolien: paysages en péril', http://ventdecolere.org/archives/doc_reference/Manifeste%20SPPEF.pdf Consulted 14.12.2006

SPPEF (2006) 'Les éoliennes: ce que nous demandons et ce nous avons obtenu', http://sppef.free.fr/texte/eoliennes_5.php Consulted 14.12.2006

Starapoli, C. (2003) 'Reforming the reform in the electricity industry: lessons from the British experience', in Glachant, J.-M. and Finon, D. (eds) *Competition in European Electricity Markets: A Cross-Country Comparison*, pp. 57–79, Cheltenham: Edward Elgar

Strachan, P. A., Lal, D. and Malmborg, F. von (2006) 'The evolving UK wind energy industry: critical policy and management aspects of the emerging research agenda', *European Environment*, 16, 1–18

Sturm, R. (1996) 'Continuity and change in the policy-making process', in Smith, G., Paterson, W. E. and Padgett, S. (eds) *Developments in German Politics 2*, pp. 117–32, Basingstoke: Macmillan

Sundquist, T. (2004) 'What causes the disparity of electricity externality estimates?', *Energy Policy*, 1753–66

Sustainable Development Commission (2005) 'Wind power in the UK', http//:www.sd-commission.org.uk Consulted 1.6.2005

Szarka, J. (2004) 'Wind power, discourse coalitions and climate change: breaking the stalemate?', *European Environment*, 14: 6 (November–December 2004), 317–30

Szarka, J. (2006) 'Wind power, policy learning and paradigm change', *Energy Policy*, 34: 17 (November), 3041–8

Szarka, J. (2007) 'Why is there no wind rush in France?', *European Environment*, forthcoming

Szarka, J. and Blühdorn, I. (2006) *Wind Power in Britain and Germany: Explaining Contrasting Development Paths*, London: Anglo-German Foundation for the Study of Industrial Society

Taylor, R. H. (1983) *Alternative Energy Sources for the Centralised Generation of Electricity*, Bristol: Adam Hilger Ltd

Technology Review (2006) 'New technologies in Spain: wind power', http://www.technologyreview.com/microsites/spain/wind/docs/TR_windpower_spain. pdf Consulted 6 June 2006

Thayer, R. L. and Freeman, C. M. (1987) 'Altamont: public perceptions of a wind energy landscape', *Landscape and Urban Planning*, 14: 379–98

Tinkerman, R. (2006) 'Engel, Vestas and the will to wind', *Windpower Monthly*, 22: 7 (July), 6

Toke, D. (2002) 'Wind power in the UK and Denmark: can rational choice help explain different outcomes?', *Environmental Politics*, 11: 4 (Winter), 83–100

Toke, D. (2005a) 'Will the government catch the wind?', *The Political Quarterly*, 76: 1 (January), 48–56

Toke, D. (2005b) 'Explaining wind power planning outcomes: some findings from a study in England and Wales', *Energy Policy*, 33: 12 (August), 1527–39

Toke, D. (2005c) 'Community wind power in Europe and in the UK', *Wind Engineering*, 29: 3, 301–8

Toke, D. and Strachan, P. A. (2006) 'Ecological modernisation and wind power in the UK', *European Environment*, 16, 155–66

Tranæs, F. (1996) 'Danish wind energy', http://www.dkvind.dk Consulted 12.8.2004

Van Est, R. (1999) *Winds of Change. A Comparative Study of the Politics of Wind Energy Innovation in California and Denmark*, Utrecht: International Books

VdC (2004) 'Manifeste', http://ventdecolere.org/archives/doc_reference/ Manifeste 2004.pdf Consulted 14.12.2006

VdC (2005) 'Vademecum: que faire pour s'opposer à l'implantation d'éoliennes près de chez soi?', http://ventdecolere.org/archives/doc_reference/VADEME-CUM.pdf Consulted 14.12.2006

VdC (2006) 'L'arnaque de l'éolien industriel français', http://ventdecolere. org/archives/doc_reference/ARNAQUE-EOLIEN-INDUSTRIEL.pdf Consulted 19.6.2006

VDEW (2005) 'Proposal for a discussion on how to promote renewable energies in the future: achieving extension targets efficiently', Berlin: VDEW

VDMA (2005) 'German wind industry: market development and perspectives', Berlin: VDMA

Viertl, C. and Bömer, J. (2005) 'Development of German policies on offshore wind energy', Berlin: BMU

VoS (2003a) 'Wind power and the planning system', Briefing paper Vol. 1, No. 1 (November) http://www.viewsofscotland.org/VoS_Briefs.htm Consulted 18.7.2005

VoS (2003b) 'The ROC scam', Briefing paper Vol. 1, No. 2 (November) http://www.viewsofscotland.org/library/docs/VoSB_The_ROC_Scam.pdf Consulted 7.12.2006

VoS (2004) 'A walk in the dark', Briefing paper Vol. 1, No. 5 (March)

VoS (2004b) 'Vos News', Vol. 2, No. 2 (October) http://www.viewsofscotland.org/ VoS_Briefs.htm Consulted 18.7.2005

VoS (2005) 'US wind chiefs pull funding slur' in *VoS News*, Vol. 2, No. 4 (March), 24

VoS (2006) 'Who we are and what we do', http://www.viewsofscotland.org/about/ Consulted 7.12.2006

Waller-Hunter, J. (2004) 'The contribution of renewable energies in meeting the climate challenge', Keynote Address at the International Conference for Renewable Energies, Bonn, Germany, 1–4 June 2004, http://ttclear.unfccc.int/ttclear/pdf/News/jwh040604.pdf Consulted 10.2.2005

Weizsäcker, E. von, Lovins, A. B. and Lovins, L. H. (1997) *Factor Four: Doubling Wealth, Halving Resource Use*, London: Earthscan

Witherspoon, S. (1994) 'The greening of Britain: romance and rationality', in Jowell, R. *et al.* (eds) *British Social Attitudes Survey. The Eleventh Report*, pp. 107–39, Aldershot: Gower

World Commission on Environment and Development / Brundtland Report (1987) *Our Common Future*, Oxford: OUP

Wurzel, R. K. W. (2002) *Environmental Policy-Making in Britain, Germany and the European Union*, Manchester: Manchester University Press

Wüstenhagen, R. and Bilharz, M. (2005) 'Green energy market development in Germany: effective public policy and emerging customer demand', *Energy Policy*, in press

Yes2Wind (2004) 'Debunking the myths', http://www.yes2wind.com/debunk.html Consulted 14.12.2006

Young, A. (2001) *The Politics of Regulation. Privatised Utilities in Britain*, Basingstoke: Palgrave

Zervos, A. (2003) 'Developing wind energy to meet Kyoto targets in the European Union', *Wind Energy*, 6: 3, 309–19

Index

Acciona, 26, 34, 40, 45
ADEME, 49
AEE, 47
Africa, 7
Altamont, 40, 164, 170
alternative energy movement, 4–6,
    19, 22, 30–1, 36, 41, 42, 46, 72,
    183, 193 *see also* NGOs
Andalusia, 127, 141, 203
antinuclear movement, 18, 30–1, 32,
    46, 48–9, 116, 165, 168, 180 *see
    also* NGOs
antiwind groups, 19, 21, 59, 83,
    162–9, 175, 176, 191, 192
  Country Guardian, 163, 165, 180,
    200
  Neighbours against Windmills, 162
  Renewable Energy Foundation, 163
  Vent de colère, 163, 180
  Views of Scotland, 163
  see also NGOs
APPA, 47
Aragón, 127, 141, 149, 157
Areva, 27, 55, 183
Asturias, 149
Austria, 82
Average Reference Tariff, 78

Babcock Brown, 183
Baden-Württemberg, 141
balancing services, 85, 98, 99, 100,
    108, 120, 121, 122, 123, 124
Barningham High Moor, 203
Bavaria, 141
Baywind, 35, 202
Belgium, 118
Blacklaw, 151, 200
Blair, Tony, 55, 83, 85
BNA, 15, 203
Bonus, 25, 26
Brandenburg, 32, 147
British Gas, 84
Brittany, 142

bulk energy, 7, 194
bulk power, 5, 6, 19, 22, 36, 41, 42,
    136, 183, 193, 194, 197, 199
BWE, 47
BWEA, 47, 57, 58, 59, 92, 125, 150,
    170, 180

Canary Islands, 203
capitalism,
  large-scale, 22, 29, 34, 35, 36, 44–5,
    91–3, 137, 182, 183, 193
  small-scale, 22, 32, 36, 91–3, 182,
    183, 185, 193
carbon capture and storage, 115, 129,
    134
Carbon Trust, 49, 134
Carland Cross, 37
Castilla-La Mancha, 141, 148, 157
Castilla-Leon, 141, 157
Catalonia, 127
Causeymire, 165
Cefn Croes, 200, 203
CEGB, 6
CESA, 34
China, 8, 25, 157
climate change, 1, 9, 15, 52, 53, 56,
    57, 68, 128, 130, 165, 171, 173,
    177, 179, 191
climate policy, 136, 137, 164, 177,
    178, 191
  Climate Change Levy, 135, 202
  Emissions Trading Scheme
    (EU-ETS), 63, 64, 135, 136, 188
  Emissions Trading Scheme (UK), 135
  Kyoto Protocol, 33, 52, 60, 61, 76,
    132
    flexible mechanisms, 135, 136
  United Nations Framework
    Convention on Climate
    Change, 33
  see also greenhouse gas emissions
combined heat and power, 113, 116
community trust funds, 158, 186

Copenhagen, 127
Countryside Agency, 172, 203
Countryside Commission, 172
CRE, 15, 165

De Wind, 28
Delabole, 35, 37, 142
DENA, 49, 126
Denmark, 2, 5, 7, 19, 23, 24, 25–6, 28,
    29–31, 34, 36, 40, 42, 43, 44, 48,
    56, 58, 59, 62, 66, 68–73, 78, 83,
    87, 88, 91, 92, 93, 94, 95, 99, 100,
    102, 105, 106, 108, 111, 112, 113,
    116, 117, 118, 122, 123, 125, 126,
    127, 128, 131, 140, 143, 144–7,
    152, 153, 154, 158, 159, 160, 161,
    162, 165, 179, 184, 185, 186
deregulation, 15
development paths
    choices, 2–13, 18, 21, 46, 62, 87,
        92, 110, 115, 182, 192–9
    hard path, 2, 3–6, 22, 41, 42, 45,
        128, 136, 192, 193, 194, 195
    path dependence, 13, 72, 106, 198
    soft path, 2, 3–6, 22, 41, 42, 45,
        128, 192, 193, 195
devil shift, 168–9
discourse coalitions, 57, 174
DONG, 40
DTI, 85
DWIA, 47, 70
DWTOA, 31, 47, 69, 70

ecological modernisation, 3, 10–12,
    45, 61, 76, 194
ecomodernist growth paradigm, 12,
    45, 194
economic policy, 7–9
Ecotèchnia, 26
EdF-Energies Nouvelles, 27
electricity supply industry, 1, 13, 16,
    36, 42, 54, 56, 58, 59, 63, 70, 90,
    92, 93, 105, 110–37, 183, 193, 201
electricity utilities, 14, 30, 31, 32, 33,
    34, 36, 41, 42, 44, 55, 56, 70, 72,
    74, 78, 79, 92, 93, 104, 110, 113,
    120, 124, 125, 126, 136, 146, 156,
    175, 182, 185
    British Energy, 55, 84

Compagnie générale des eaux, 118
Compagnie nationale du Rhône, 118
E.ON, 28, 29, 33, 93, 97, 119
E.ON Netz, 121, 126, 164, 201
EdF, 27, 29, 40, 55, 118
EHN, 34, 157
Elsam, 118
EnBW, 119, 201
Endesa, 34, 40, 118
Energi E2, 118
Iberdrola, 34, 40, 118, 157, 183
London Energy/EdF, 97
National Grid Transco, 117
National Power/Innogy, 117
Powergen, 84, 117
PreussenElektra, 32–3, 70, 201
RWE, 28, 29, 33, 35, 40, 93, 97,
    117, 119, 201
Schleswag, 33
Scottish and Southern Energy
    Supply, 84, 97, 117
Scottish Power, 40, 97, 117
Société lyonnaise des eaux, 118
Unión Fenosa, 118
Vattenfall, 40, 118, 119, 203
embedded generation, 5, 41, 124,
    136, 183, 184, 193
employment, 7, 13, 26, 27, 28, 29, 43,
    54, 76, 109, 114, 156–7, 158, 165,
    189, 195
Enercon, 26
energy 'decoupling', 11
energy efficiency, 2, 8, 49, 68, 77,
    130, 133
energy independence, 7, 53, 76, 77,
    79, 79, 86, 114, 115, 193
energy policy, 7, 9, 53, 64, 68, 76,
    106, 109, 128, 148, 163, 164, 180,
    189, 191
England, 44, 83, 84, 97, 99, 117, 127,
    128, 142, 143, 149, 150, 151, 152,
    172
English Nature, 57
Enron Wind, 25
environmental policy, 9–10, 76
environmental pressures, 9–10, 51–3,
    68, 76, 81, 86, 89, 113, 115, 140,
    148, 155, 163, 165, 168, 172, 174,
    176, 197, 199

equity, 12, 57, 104, 122, 159, 160, 177, 185, 186, 191, 195, 197, 198, 199
European Commission, 20, 72, 73, 105, 113, 126
European Court of Justice, 32–3, 55, 70, 71, 72
European Environment Agency, 128
European Union, 20
  Large Combustion Plant Directive, 116
  Renewables Directive, 61, 65, 66, 75, 83, 89, 94, 101, 105
EWEA, 47
externalities, 9–10, 63, 119, 176, 178, 186, 197
Extremadura, 141, 170

Falck Renewables, 28
FEE, 47
Finland, 131
fossil fuels, 7–8, 51, 54, 55, 63, 111, 112, 115, 116, 129, 130, 135, 136, 137, 184
  coal, 7, 55, 63, 111, 114, 115, 117, 129, 130, 202
    coal-fired generation, 15, 48, 54, 61, 111, 112, 113, 114, 115, 116, 121, 122, 130, 133, 134, 137, 183, 184
  gas, 54, 106, 111, 113, 116, 119, 121, 129, 130, 133, 134, 183, 193, 197
    dash-to-gas, 114
    North Sea, 28, 82, 111, 114
    offshore, 7
    price increases, 54, 79, 99, 114, 185, 186
    sources, 8, 53
  oil, 115, 193, 197
    crises, 7, 68, 113
    depletion, 8, 51, 53
    price increases, 8, 54, 63, 113, 114, 185
FPL, 40
France, 2, 7, 14, 19, 20, 23, 24, 27, 29, 35, 36, 43, 44, 49, 62, 66, 79–82, 83, 86, 88, 91, 92, 94, 95, 99, 100, 107, 111, 112, 115, 117, 118, 122, 123, 125, 127, 130, 134, 142, 143,
152–3, 154, 157, 158, 159, 160, 161, 162, 163, 184, 187, 189, 190

Galicia, 141, 148, 157
Gamesa Eólica, 26
GE Energy, 25
General Electric, 183
Germany, 2, 5, 7, 12, 19, 20, 23, 24, 25, 26, 28, 29, 32–4, 36, 43, 44, 48, 49, 55, 56, 58, 59, 62, 66, 73–6, 78, 83, 86, 87, 88, 92, 93, 94, 95, 99, 100, 102, 105, 106, 107, 111, 112, 113, 115, 116, 118, 122, 123, 125, 126, 127, 128, 131, 141, 143, 147–8, 152, 153, 157, 158, 159, 160, 161, 163, 175, 179, 184, 185, 186, 187, 189
Goldwind, 25, 102
governance, 3, 10–12, 192, 195
greenhouse gas emissions, 9, 33, 48, 52, 76, 89, 110, 115
  emissions factors, 130–5
  emissions reduction pathways, 20, 55, 60, 61, 68, 128–37, 188
  see also climate policy
green movement, 27, 31, 32, 114, 122, 137, 183 *see also* NGOs
green parties, 32, 48–9, 116
grid
  configuration, 1, 5, 14, 15, 16, 17, 119, 127–8, 136, 140, 197
  connection and access, 1, 14, 16, 43–4, 61, 67, 69, 74, 82, 99, 103, 108, 122, 125–6, 138, 141, 149, 152, 153, 160, 169, 187, 190
  integration of wind power, 20, 72, 73, 110, 120, 125–8, 136, 183
  management, 126–7, 184
  operators, 120, 122
  prioritised dispatch, 69, 71
  reinforcement, 127–8, 184
  stability, 126, 164, 184
  strong grids, 14, 125
  weak grids, 14, 16, 125
GWEC, 47, 129

Hadyard Hill, 151, 200

IDAE, 49
IEA, 49
India, 8, 25
industrial policy, 7, 27, 65, 69, 76,
    106, 109, 113, 165, 189
Ingetur, 26
intermittence, 38–9, 120–2, 124, 129,
    164, 174–5, 183, 184
Iran, 8
Iraq, 8
Ireland, 83, 117, 142, 150
Isle of Lewis, 171
Italy, 118

Jeumont, 27
Juul, Johannes, 25
Jutland, 127, 140, 157

La Cour, Poul, 25
land rents, 139, 157, 159
landfill gas, 86, 96, 100, 111
Languedoc-Roussillon, 142
large-scale penetration (of wind
    power), 55, 61, 95, 125, 133, 184,
    193
LG Glasfiber, 26, 28
liberalisation of electricity markets,
    15, 34, 59, 63, 70, 72, 74, 77, 89,
    90, 91, 105, 110, 111, 117–19,
    122, 123, 197, 198
Lower Saxony, 141

M. Torres, 26
Madrid, 141
market integration, 20, 72, 73, 105,
    110, 120, 122–5, 136, 184
Mecklenburg-Vorpommern, 141
Middle East, 7
Ministry of Defence (UK), 170, 176
Morocco, 128

National Wind Power, 28, 35
Natural England, 170, 172, 203
nature conservation, 144, 146, 147,
    148, 153, 155, 166, 167, 172, 176,
    177, 178
  bats, 170
  birds, 147, 163, 170–1, 177, 192
Navarra, 34, 141, 148, 157, 163, 170,
    203

NEG Micon, 25, 26
NETA, 123
Netherlands, 12, 25, 126
NGOs, 15, 19, 36, 52, 55, 57–8, 61,
    155, 169, 170, 178, 180, 191, 193
  Friends of the Earth, 15, 47, 48, 59,
    177, 180, 200
  Greenpeace, 15, 47, 48, 59, 177,
    180
  OOA, 31
  OVE, 31, 70
  WWF, 47, 57, 59, 177
  amenity associations, 21, 173,
    176, 177, 180
    Open Spaces Society, 173
    Ramblers' Association, 173–4
  countryside/landscape protection
    groups, 150, 155, 176, 177,
    179, 180, 192
    Bundesverband
      Landschaftschutz, 163
    CPRE, 150, 172, 177, 179
    CPRW, 172, 177
    SPPEF, 172
    Europa Nostra, 172
  nature conservation groups, 21,
    170, 177, 179
    Danmarks
      Naturfredningsforening,
      155
    RSPB, 57, 170–1, 177
    SEO/Birdlife, 177
  see also prowind groups, antiwind
    groups
Nordex, 25, 26
Norway, 79, 82, 117, 127, 130
nuclear power, 4, 5, 7, 9, 15, 18, 27,
    31, 32, 48, 51, 53, 54, 55, 56, 59,
    61, 63, 79, 82, 83, 106, 111, 112,
    113, 114, 115, 116, 117, 118, 120,
    121, 122, 130, 134, 137, 165, 183,
    184, 197, 203
  European Pressurised Reactor, 116

Ofgem, 15, 100
offshore (wind power), 42–4, 201, 203
  Barrow, 44
  Borkum West, 44
  Horns Rev, 42, 44

offshore (wind power) – *continued*
Mittelgrunden, 43
Nysted, 42
Thames Estuary Array, 42
Tunø Knob, 42
Vindeby, 42
ownership (of wind installations), 19,
29–36, 89, 91–2, 99, 124, 146,
153, 155–6, 158, 159, 160, 182
Bürgerwindparks, 32
community/local, 5, 6, 19, 32, 35,
36, 41, 56, 58, 92, 138, 146,
154, 155, 161, 182, 183, 185,
190, 193
cooperatives, 30, 32, 35, 36, 92,
155, 185, 202
'Danish model', 30, 34, 35, 36, 40,
44, 142, 182, 183, 193
farmers, 5, 30, 32, 92, 143, 146,
157
'international utility model', 36, 40,
42, 44, 183, 193
'Spanish model', 34–5, 36, 44, 182
utility owned, 34, 69, 156, 159,
161, 162

PER, 127, 149
Planning (land use), 15, 16, 20, 59,
61, 72, 82, 128, 138, 139–40,
143–54, 159, 166, 167, 169, 173,
176, 178, 190, 203
criteria based planning, 149, 151,
152, 154, 155, 159, 167, 190
NPPG6, 150
PPG22, 150
PPS22, 151, 154, 190
spatial planning, 145, 146, 147,
148, 151, 152, 153, 155, 158,
159, 190
statutory consultees, 170, 176
strategic locational guidance, 138,
145, 154, 160, 171, 190, 192
TAN8, 151
see also siting issues
Poland, 126
policy design (in relation to wind
power), 2, 20, 64–7, 68–72, 73–5,
77–9, 80–2, 88–93, 96, 98, 101,
174, 175, 185, 186, 189

capacity planning (indicative
planning), 68, 76, 79, 87,
89–91, 106, 190
cost-reflective subsidies, 100–1, 104,
106, 109
degression, 75, 79, 80, 82, 87, 99,
100, 101, 104
effectiveness, 89, 93–8, 105, 108
efficiency, 89, 93, 94, 98–102, 105,
108
feed-in tariffs, 32, 35, 67, 69–72,
73–6, 77–9, 80–2, 84, 86, 87,
88–109, 102–5, 107–9, 122,
123, 124, 152, 153, 164, 185,
187, 189
German feed-in laws, 73–4, 80
harmonisation of policy at EU level,
20, 73, 105, 108, 188
purchase obligation, 74, 77, 80, 85,
104, 123
quota systems, 67, 70–2, 88–109,
124
Renewables Obligation, 83–6, 89,
92, 93, 96–104, 123, 135,
150, 165, 175, 186, 189
Spanish 'special regime', 77–9, 201
tender schemes, 67, 81, 88, 92, 161
France – *Eole 2005*, 79–80
UK – NFFO, 80, 82–3, 201
policy frames, 2–3, 6–10, 51–6, 192–3
policy learning, 20, 73, 91, 88–109,
143, 153, 183, 184–92
policy recommendations, 21, 105–9,
182, 187–92
population density, 140, 141, 143,
159
Portugal, 28, 128
power purchase agreements, 85, 97–8,
123, 124
Powergen Renewables, 28
privatisation, 15, 117
production forecasts, 124
prowind groups
'Embrace the revolution', 57–8, 180
Yes2Wind, 48, 58–9, 180
see also NGOs
public ownership, 14, 117, 119

Quebec, 200

REE, 118, 127
REFITs, *see* Policy design – feed-in tariffs
regulation, 15, 17, 67, 90, 103, 107–8, 154, 156
regulatory competition, 70, 105
Rejsby Hede, 40, 146
renewable energy sources, 2–5, 7–10, 12, 20–2, 26–8, 33, 47–51, 53–6, 60–8, 70, 73–84, 86–91, 94–9, 101, 105–8, 111–12, 114–17, 119, 120, 122, 123–5, 128, 130, 132, 134, 136, 140, 145, 150, 151, 157, 158, 161, 163–5, 171–3, 175, 180, 182–4, 187–9, 191, 192, 194–9
   biomass, 8, 95, 111
   co-firing of biomass, 96, 202
   hydroelectric power, 9, 10, 27, 35, 53, 66, 76, 79, 81, 82, 90, 95, 96, 101, 111, 112, 114, 115, 120, 121, 130, 183, 184
   marine renewables, 9, 17, 43, 86, 189
   photovoltaic (PV), 74, 90, 101, 175
Renewable Energy Systems, 28
Renewables Obligation *see* Policy design – quota systems
REpower, 26, 27
repowering, 71, 75, 145, 200, 201
Rio world conference and declaration, 11
Risø research institute, 72
RPS, *see* Policy design – quota systems
Russia, 7, 113

Saxony-Anhalt, 32, 141
Schleswig-Holstein, 32, 141, 147
Scotland, 83, 84, 117, 97, 99, 127, 128, 142, 143, 149, 150, 151, 152, 158, 171
Scottish Executive, 150, 152
Scottish Natural Heritage, 170, 171
Scottish Renewables Forum, 49
security of supply, 7, 13, 14, 51, 53, 68, 74, 83, 86, 89, 105, 113, 114, 137, 193, 195, 197
self-sufficiency (in energy sourcing), *see* energy independence
SER, 47
Shell, 40, 183
Siemens, 28, 183

SIIF-Energies, 27
siting issues, 20, 139–40
   cumulative effect, 125, 146, 148, 159, 171, 190
   electromagnetic interference, 144, 170, 175
   environmental impact assessment, 144, 153, 166
   heritage sites, 153, 171, 172, 192
   housing, 140, 162–3, 166, 167, 168, 172
   landscape value, 1, 16, 83, 107, 128, 139, 142–3, 144, 145, 146, 149, 150, 153, 162, 166, 171, 172, 173, 177, 179
   noise, 144, 146, 148, 156, 162, 163, 166, 186
   radar, 170, 175
   regional concentration, 20, 80, 107, 125, 138, 140–3, 159
   saturation, 140, 148, 149, 153, 154, 159, 190
   visual impact, 1, 43, 144, 148, 150, 156, 162, 166, 186
   see also planning, wind speeds
social acceptability, 17–18, 29, 138, 143, 149, 159, 160, 168, 169, 170, 172–4, 181, 182, 192, 195–6
social acceptance, 17, 20, 29, 36, 42, 92, 138, 140, 145, 147, 154–8, 159, 183, 190, 195
social contracts, 13–17, 21, 31, 58, 146, 160, 182, 186, 190, 195–9
social costs, 9, 163, 176, 186, 197
Spain, 2, 7, 19, 20, 23, 24, 25, 26–7, 28, 28, 29, 34–5, 36, 40, 43, 44, 49, 62, 66, 76–9, 83, 86, 87, 88, 91, 92, 93, 94, 95, 100, 102, 105, 106, 107, 111, 112, 114, 116, 117, 118, 119, 122, 124, 125, 126, 127, 130, 141, 143, 148–9, 154, 157, 158, 159, 160, 161, 184, 185, 186
story-lines, 50, 191
   of antiwind protest, 174–9
   of the wind lobby, 50–6
sustainable development, 1, 3, 5, 10–12, 21, 36, 45, 51, 57, 68, 89, 173, 177, 178, 182, 194, 198, 199, 200

Sustainable Development
  Commission, 49, 200
Suzlon, 25, 102
Sweden, 127
Switzerland, 118

Tarifa, 141, 148, 163, 170
taxation, 157, 158–9
technology choice, 6, 31, 66, 164,
  169, 198
technology differentiation, 74–5, 104,
  106, 109, 187
Thatcher, Margaret, 114, 117, 200

utilities, *see* electricity utilities
UK, 2, 7, 19, 20, 23, 24, 28–9, 33, 35,
  36, 39, 40, 43, 44, 49, 58, 62, 66,
  82–6, 88, 92, 93, 94, 95, 96–103,
  111, 112, 114, 115, 117, 118,
  122, 123, 125, 126, 127, 130,
  132, 133, 142, 143, 149–52, 153,
  154, 158, 159, 160, 161, 162,
  163, 167, 175, 176, 179, 184,
  185, 187, 189, 190
USA, 23, 25, 39, 40

Valencia, 127, 171
VDMA, 26, 47, 189, 201
Vergnet, 27
Vestas, 25, 26, 28, 44, 157

Wales, 83, 84, 99, 117, 142, 143, 149,
  150, 151, 152, 158, 173
Welsh Assembly, 151
Westmill Wind Farm, 202
Whinash, 172, 203
Whitehall, 152
Whitelee, 151, 200
wind farms
  categories, 39–40
  concept, 146
wind forecasts, 203
wind lobby, 19, 46–50, 130, 167, 168,
  169
wind power capacity levels, 22–4
wind power deployment patterns,
  36–44 *see also* siting issues
wind production schedules, 78
wind speeds, 17, 37, 38–9, 73, 75, 80,
  99, 107, 120, 139, 140, 141, 142,
  159, 179, 187
Wind Supply project, 28
wind turbines
  'Danish concept', 1, 25
  exports, 28, 29, 54, 69, 102, 109,
    165, 182, 189
  Gedser turbine, 25
  GROWIAN project, 73
  physical scale, 1, 37–40
  wind turbine industry, 22–9
Wind World, 25
Windprospect, 28